The Human Relationship with Nature

The Human Relationship with Nature
Development and Culture

Peter H. Kahn, Jr.

The MIT Press
Cambridge, Massachusetts
London, England

This book was set in Sabon by Wellington Graphics.

Printed and bound in the United States of America.

Library of Congress Cataloging-in-Publication Data

Kahn, Peter H.
  The human relationship with nature : development and culture /
  Peter H. Kahn, Jr.
    p. cm.
  Includes bibliographical references and index.
  ISBN 0-262-11240-X (hardcover : alk. paper)
  1. Nature—Psychological aspects. 2. Nature—Psychological
aspects—Cross-cultural studies. 3. Environmental psychology.
I. Title.
BF353.5.N37K34 1999
155.9′1—dc21   98-47085
                                                                        CIP

To my mother,
Molly Kahn

Contents

Acknowledgments

Over the years my ideas have been challenged and supported—and often the latter by means of the former—by numerous people whom I have had the good fortune to know. Allen Black, Jonas Langer, and Samuel Scheffler provided early mentoring. Rheta DeVries heartened my scholarly pursuits. James L. Jarrett provided breath and vigor. Elliot Turiel grounded my scholarship, and continues to guide, sometimes by what he does not say, always by example.

In each of the five empirical studies reported in this book, I have benefited greatly from the collaboration of colleagues and students alike. I would like here to mention the contributions of specific individuals, and to offer them my heartfelt thanks.

The Houston child study. Batya Friedman was my colleague. Ann McCoy assisted with data collection, and Tracey Hardman and Daniel Howe assisted with analysis and coding.

The Houston parent study. Batya Friedman was my colleague. Ann McCoy assisted with data collection, and Daniel Howe and Sasha Cornell assisted with analysis and coding.

The Prince William Sound study. Ann McCoy, Kiki Przewlocki, and David Siegel assisted with data collection. Tracey Hardman and Daniel Howe assisted with analysis and coding.

The Brazilian Amazon study. Daniel Howe and Batya Friedman were my colleagues. Carlos Miller (and the staff of the Fundação Victória Amazônica) and Miguel Rocha da Silva of Amazon Nut Safaris assisted with arrangements in Brazil. Marilena Gouvêa translated the interviews from Portuguese into English. Sue Nackoney assisted with coding.

The Portugal study. Orlando Lourenço was my colleague. The 1996 and 1997 classes in developmental psychology at the University of

Lisbon, Portugal (under the direction of Orlando Lourenço) assisted with data collection and transcriptions (in Portuguese). Marilena Gouvêa translated the interviews from Portuguese into English. Todd Covert assisted with data analysis and coding.

In each study, Sara Brose has assisted wonderfully with the statistical analyses.

Friends and colleagues have generously commented on portions of earlier drafts of this book: Robert Campbell, John Coley, Barbara Dean, Batya Friedman, Charles Helwig, Stephen Kellert, Orlando Lourenço, and Patricia Nevers. Amy Brand and Clay Morgan of The MIT Press have provided editorial assistance. I also extend my appreciation to Paul Ammon, William Arsenio, Seth Benz (and students and other faculty of the Audubon Expedition Institute), Angela Biaggio, Clifton Bowen, Carolyn Hildebrandt, Willet Kempton, Melanie Killen, Nadine Lambert, Marta Laupa, Terry Madden, Leonard Marascuilo, Gene Myers, Nancy Nordmann, Larry Nucci, John Ogbu, William Rohwer, Herbert Saltzstein, Carol Saunders, Judith Smetana, Mitchell Thomashow, and Cecilia Wainryb.

In pursuing my research, I have been appreciative of grants from the Spencer Foundation, Texas Education Agency, Colby College (Interdisciplinary Studies Grant), and the University of Houston (Limited Grant-in-Aid).

Some of the material in this book draws on my published papers listed below. The material is reprinted with permission from the publishers.

Howe, D., Kahn, P. H., Jr., and Friedman, B. (1996). Along the Rio Negro: Brazilian children's environmental views and values. *Developmental Psychology* 32, 979–987 (American Psychological Association).

Kahn, P. H., Jr. (1991). Bounding the controversies: Foundational issues in the study of moral development. *Human Development* 34, 325–340 (Karger, Basel).

Kahn, P. H., Jr. (1992). Children's obligatory and discretionary moral judgments. *Child Development* 63, 416–430 (Society for Research in Child Development).

Kahn, P. H., Jr. (1995). Commentary on D. Moshman's "The construction of moral rationality." *Human Development* 38, 282–288 (Karger, Basel).

Kahn, P. H., Jr. (1997a). Developmental psychology and the biophilia hypothesis: Children's affiliation with nature. *Developmental Review* 17, 1–61 (Academic Press).

Kahn, P. H., Jr. (1997b). Children's moral and ecological reasoning about the Prince William Sound oil spill. *Developmental Psychology* 33, 1091–1096 (American Psychological Association).

Kahn, P. H., Jr. (1997c). Bayous and jungle rivers: Cross-cultural perspectives on children's environmental moral reasoning. In H. Saltzstein (Ed.), *Culture as a context for moral development: New perspectives on the particular and the universal.* New Directions for Child Development (W. Damon, Series Editor) (pp. 23–36). San Francisco, CA: Jossey-Bass.

Kahn, P. H., Jr., and Friedman, B. (1995). Environmental views and values of children in an inner-city Black community. *Child Development* 66, 1403–1417 (Society for Research in Child Development).

Kahn, P. H., Jr., and Friedman, B. (1998). On nature and environmental education: Black parents speak from the inner city. *Environmental Education Research* 4, 25–39 (Carfax Publishing Limited).

Preface

Here on these 670 acres of mountain meadows and forest I have found that the less I import, the more I uncover. My wife, daughter, and I share this land with twelve other families, and we live here when we can, which is usually four months of the year and sometimes longer.

One day last December I could be found swimming in the nearby river. With the current rains this river was big and swift. Two men, T. A. and Jonas, were swimming with me. We had begun the day in T. A.'s cabin asking questions about our relationship with the land. Could we be in harmony with the land and still cut trees for our building projects? How many trees? Jonas says you can tell by how it feels at the time. I remind Jonas that another member of our community says it would feel right to have our land commercially logged. Jonas says that is not a true feeling. I ask, How does one know a true feeling? Do not our feelings about our relationship with the land depend on our thoughts, beliefs, and values? In turn, where do they come from? T. A. and Jonas also tempt me with an idea. They are interested in purchasing a Woodmizer, a portable mill with which we could cut our own timber into lumber. They know I like to sink my chain saw into logs, and feel its bite, and transform nature into something for human use: a barn, firewood, furniture. "Think what you could do with a Woodmizer," they say. That is the scary part, for the same impetus by which I use nature in seemingly wholesome ways can lead me to walk our land and see not trees but lumber, and not meadows but housing sites. Does this impetus have biological roots? And within Western culture in particular has this impetus—in consort with our increasing technology—become unleashed from natural rhythms and checks?

We keep asking such questions. Finally, we walk out into the rain and run across meadows and down a steep forested trail to the river. That is when we could be found swimming, by which I mean we get into the water and dunk our heads and go with the current for about ten feet and get out. That fast. Then, of all things, the rain feels warm and soothing against my skin. For twenty–five years I have swum in this river with these men who, back then, were boys. We have a history together, and a future. Later, on parting ways in a meadow, I tell T. A. and Jonas that we had partially answered our earlier questions. "Yes?" Jonas asks. "The swim," I say. The direct experience of the land helps ground our being and our relationships with one another.

They know that. But they also know that more needs to be said. And here, in this book, I pick up the conversation more formally.

Introduction

Over the past eight years my colleagues and I have conducted research on how humans develop a relationship with nature. We have chosen diverse geographical locations, ranging from an economically impoverished black community in Houston, Texas, to a remote village in the Brazilian Amazon. Our participants have comprised six-year-old children through young adults. Through our research we have sought answers to the following questions. How do people value nature and morally reason about its preservation? Do children have a deep connection to the natural world, which in time gets largely severed by modern society? Or do such connections emerge, if at all, in adolescence or later, and perhaps require increased cognitive capacities and moral sensibilities? Are people's environmental values and reasoning mentally organized (structured), and do such structures develop such that our societal discourse on environmental issues has its genesis in childhood? How does culture affect environmental commitments and sensibilities? Are there universal features in the human relationship with nature?

This topic of the human relationship with nature can be called big in scope, interdisciplinary. It involves understanding our biological roots, which I shall pursue. It also involves understanding environmental behavior, history, policy, and science, and issues across many other fields today. But my focus shall be less on these issues and more on understanding the development of our environmental moral reasoning and values. For such understandings, when attained, capture that which is at once deeply fundamental to our being and very practical. In his classic essay on the conservation ethic, Leopold ([1949] 1970) writes of his disappointment with the slow progress in conservation education, and that the "usual answer to this dilemma is 'more conservation education.'" In turn, he argues that such education will continue to fail until we help people develop a "love, respect, and admiration for land, and a high regard for its value" (p. 261). "No important change in ethics," Leopold writes, "was ever accomplished without an internal change in our intellectual emphasis, loyalties, affections, and convictions" (p. 246). That is much of what I am after: understanding, with respect to nature, our intellectual emphasis, loyalties, affections, and convictions.

I have written this book for social scientists in general, and psychologists in particular, who are interested in related lines of inquiry. Many fields can be strengthened by the theoretical grounding, methodology, and results offered herein. I have also written this book for an informed generalist interested in environmental issues and children. It is an accessible scholarly treatise, which I trust is not a contradiction in terms.

In chapter 1, I bring together much of the promising research that supports and fleshes out "biophilia"—a term coined by Wilson (1984) to refer to what he and his colleagues hypothesize is a fundamental, genetically based human need and propensity to affiliate with life. The biophilic instinct often emerges, according to Wilson (1984), unconsciously, in our cognition, emotions, art, and ethics, and unfolds "in the predictable fantasies and responses of individuals from early childhood onward. It cascades into repetitive patterns of culture across most or all societies" (p. 85). For empirical support, note that recent studies have shown that even minimal connection with nature—such as looking at it through a window—increases productivity and health in the work place, promotes healing of patients in hospitals, and reduces the frequency of sickness in prisons. Other studies have begun to show that when given the option, humans choose landscapes that fit patterns laid down deep in human history on the savannas of East Africa. Direct contact with animals has been shown to greatly benefit a wide range of clinical patients: from adults with Alzheimer's disease to autistic children. Wilson (1992) points out that people crowd national parks to experience natural landscapes, and "travel long distances to stroll along the seashore, for reasons they can't put into words" (p. 350). Indeed it would appear more than a cultural convention that flowers are often sent to sick people in a hospital, or during time of mourning. The need and propensity to affiliate with nature appears great. Moreover, if biophilia has merit, it helps establish normative principles. The logical move would take the form of if-then statements such as, "If you want to promote your physical and psychological health, then affiliate with nature." Or, more broadly, "If we want a healthy society, then we need to preserve nature so that we still have something natural with which to affiliate."

But does the biophilia hypothesis have merit? After all, it is not difficult to generate examples that would offer seemingly disconfirming evidence. We probably all know people, for example, who appear to dislike nature. Conversely, most people appear to enjoy human artifacts, but is that cause to coin a term like "artifactphilia"? Or, if biophilia has a genetic base, how do its proponents explain the seemingly rapid conversion of many native people to a Western, and presumably less biophilic, culture? These questions, and many others, have been at the forefront of current debate. In chapter 2, I provide some answers.

In answering, however, I cast biophilia in somewhat different ways than Wilson and others have. My reasoning is that by grounding bio-philia in evolutionary biology, Wilson opens up controversies that engage classic issues going back to the 1970s on the adequacy of sociobiological theory to explain human behavior. Here I suggest that functional evolutionary accounts (biology, genes, and genetic fitness) sometimes constrain the biophilia construct, and that the human affiliation with nature needs to be investigated in ways that take development and culture seriously. Yet, to speak plainly, I do not think most evolutionarily inclined theorists understand how their psychological theories limit them from doing so. Namely, as I explain in chapter 2, most theorists working within the evolutionary framework hold to a mechanistic conception of the human mind. Cosmides, Tooby, and Barkow (1992), for example, write of "information-processing mechanisms situated in human minds" (p. 3), and of the brain as "a computer made out of organic compounds" (p. 8). Pinker (1997) argues "that the mind is a system of organs of computation designed by natural selection to solve the problems faced by our evolutionary ancestors in their foraging way of life" (p. x). Dawkins (1976) says: "We are survival machines—robot vehicles blindly programmed to preserve the selfish molecules known as genes" (p. ix).

Of course at times it may be useful to think of the mind as if it were a piece of machinery. But we should not lose sight of a clear distinction. As Searle (1990) notes, for certain purposes it can be useful to model water molecules with ping-pong balls in a bathtub, but no one should confuse those balls with actual water. Similarly, we should not confuse mind with machine. The issue, then, is what sort of psychology can best support concepts of intentionality, free will, and meaning, and the possibility for individuals to shape—from an ethical stance—cultural practices? Here I think structural-developmental theory has much to offer, and can direct productive research programs on understanding the human relationship with nature.

In chapter 3, I explain the theoretical assumptions and perspectives of structural-developmental theory. Briefly, this theory posits that learning involves neither simply the replacement of one view (the incorrect one) with another (the presumed correct one), nor simply the stacking, like building blocks, of new knowledge on top of old knowledge, but rather transformations of knowledge. Transformations, in turn, occur through the child's active, original thinking. As Baldwin ([1897] 1973) says, a child's knowledge "at each new plane is also a real invention. . . . He makes it; he gets it for himself by his own action; he achieves, invents it." Or, as Dewey ([1916] 1966) says: "We sometimes talk as if 'original research' were a . . . prerogative of scientists or at least of advanced students. But all thinking is research, and all research is native, original, with him who carries it on, even if everybody else in the world already is sure of what he is still looking for" (p. 148). Or think of it this way. On a daily level, children encounter problems, of all sorts: logical, mathematical, physical, social, ethical, environmental. Problems require solutions. The disequilibrated state is not a comfortable one. Thus the child strives toward a more comprehensive, more adequate means of resolving problems, of synthesizing disparate ideas, of making sense of the world.

Most of our environmental actions have moral import. We pollute air and water, deplete soil, deforest, create toxic wastes, and through human activity extinguish over 27,000 species each year (a conservative estimate). At a minimum, human welfare and justice is at stake. As one of the black parents we interviewed in Houston said, while speaking to us about air pollution:

[The air] *stinks, 'cause I laid up in the bed the other night. Kept smelling something, knew it wasn't in my house, 'cause I try to keep everything clean. Went to the window and it almost knocked me out. The scent was coming from outdoors into the inside and I didn't know where it was coming from. . . . Now, who'd want to walk around smelling that all the time?*

Indeed an increasing literature shows that politically disenfranchised people (often of color) shoulder more than their share of environmental harm.

But is nature only a medium by which human action morally affects other humans, or does nature itself have moral standing? Over the last few decades, this question has received a good deal of philosophical attention, and most environmental philosophers have in some form and to some degree sought to grant nature moral standing. Virtually no consideration, however, has been given to the question of whether children view their relationship with nature in moral terms, and, if so, whether they, too, grant nature moral standing. Do children, for example, believe that nature has rights or an intrinsic value, and, if so, how do they support such claims? To investigate these issues empirically propels us into moral psychology. In chapter 4, I build on the general psychological framework presented in chapter 3 and offer a moral developmental framework for investigating the human relationship with nature.

In chapter 5, I present my research methodology, which entails the semistructured interview and has its roots to Jean Piaget. This chapter is important for two reasons. First, to enhance the readability of each of the empirical chapters I have avoided the sort of detailed methodological presentation that in a journal article would normally precede each study's results. Thus this chapter provides the reader with the necessary methodological background to understand and assess the subsequent empirical research. Second, and equally important, I believe this methodology has far-reaching applications for others interested in pursuing their own related lines of inquiry. Thus this chapter provides researchers from other fields with enough detail to draw on these methods themselves. Topics include constructing the interview, enlisting subjects, interviewing, generating a coding manual, coding the data, reliability coding, and data analysis.

With chapter 6, we move to the heart of the book, the empirical research. I start with a study on the environmental views and values of children in an economically impoverished community in Houston.

Chapter 7 reports on a study conducted with these children's parents. The parents speak on the place of nature in their lives and on the importance of environmental education for their children. Chapter 8 focuses on children's moral and ecological reasoning about a real-life environmental disaster—the Prince William Sound oil spill, which occurred in 1989. I then extend the research cross-culturally—in chapter 9 to the Brazilian Amazon, and in chapter 10 to Lisbon, Portugal.

Throughout this research I am striving for psychological data that bear not only on ontogeny but also on larger environmental issues of import. Let me take a moment to illustrate what I mean. Houston is one of the more environmentally polluted cities in the United States. Local oil refineries contribute not only to the city's air pollution but also to distinct oil smells during many of the days. Bayous can be thought of more as sewage transportation channels than fresh water rivers. Within the community where we conducted the research, garbage was commonly found alongside the bayou and on the streets and sidewalks. In this context, colleagues and I systematically investigated children's knowledge of three different types of pollution: water pollution, air pollution, and garbage. For each type of pollution, we assessed whether children who understood about the idea of the pollution in general also understood that they directly encountered such pollution in Houston. The findings showed a consistent statistically supported pattern. About two-thirds of the children understood in general about problems of water pollution, air pollution, and garbage. However, contrary to our expectations, only one-third of the children believed that these environmental problems affected them directly.

How could this be? How could children who know about pollution in general, and live in a polluted city, be unaware of their own city's pollution? One possible answer is that to understand the idea of pollution one needs to compare existing polluted states to those that are less polluted. In other words, if one's only experience is with a certain amount of pollution, then that amount becomes not pollution but the norm against which more polluted states are measured. If this answer is correct, then it would speak to the importance of keeping environmental preserves, refuges, and parks close to (and even within) cities, and of providing means for children to experience these areas. I come back to this issue in chapter 12 on environmental education.

Indeed, what we perceived in the children we interviewed in Houston might well be the same sort of psychological phenomenon that affects us all from generation to generation—what I call "environmental generational amnesia." People may take the natural environment they encounter during childhood as the norm against which to measure environmental degradation later in their life. The crux here is that with each ensuing generation, the amount of environmental degradation increases, but each generation takes that amount as the norm, as the nondegraded condition. The upside is that each generation starts afresh, unencumbered mentally by the environmental mistakes and misdeeds of previous generations. The downside is that each of us can have difficulty understanding in a direct, experiential way that nature as experienced in our childhood is not the norm, but already environmentally degraded. Thus if environmental generational amnesia exists, it helps provide a psychological account of how our world has moved toward its environmentally precarious state.

Following the empirical chapters, I make clear in chapter 11 how I understand the epistemological status of nature and research. In doing so, I enter the often heated controversies about postmodernity, a theoretical orientation that increasingly troubles me as it becomes increasingly accepted by people whose work I otherwise respect. I argue that postmodern theory runs too rough over such concepts as rationality, truth, objectivity, logic, morality, and nature. Imagine, for example, that a fist-sized rock falls on your head. Is the rock, as some postmodernists maintain, only a social construction? A linguistic turn? A ploy to be used by people in power to oppress others? A co-constructed metaphor by which you create the rock and the rock creates you? The ideas can become ludicrous, especially for readers outside of the postmodern field. If a rock falls on your head, it will hurt, whether you are a Western urbanite or Samoan native. More substantively, I argue that nature is not a mere cultural convention or artifact, as some postmodernists maintain, but part of a physical and biological reality that bounds children's cognition. Accordingly, one should be able to find meaningful similarities in people's psychological structures across cultures. I develop this line of reasoning conceptually, integrating some of my earlier empirical data.

In the last chapter, I take the structural-developmental theoretical framework that has guided my research throughout this book and show how it can make important contributions to the field of environmental education. I frame my discourse in terms of constructivist education—wherein constructivism can be understood as the flip side of structural development—the side that highlights the developmental processes of most import to educators. The constructivist approach to education has continued to gain popularity in the educational literature. But through its popularization its meaning has often become diluted. Thus, building on work by DeVries (DeVries 1997; DeVries and Kohlberg 1990), I provide specificity for this term by discussing four shifts that move teaching from traditional to constructivist: from instruction to construction, from reinforcement to interest, from obedience to autonomy, and from coercion to cooperation. Then I discuss specific ways of implementing constructivist environmental education across a diversity of contexts: in the home, in the field (in terms of experiential education), in traditional K–12 classrooms, and in urban settings.

From a constructivist perspective, education is not so different from research. Both form part of a larger intellectual enterprise that is at the heart of this book. Imagine we are walking into unfamiliar territory. We have a choice, to follow the trails or not. Here is what happens to me when I choose the trails. I cover lots of ground fast. But I also become surprisingly inattentive to the surroundings. I start thinking about past experiences and future possibilities, and as my mind chatters to itself time goes by and miles are covered. In contrast, when I move off trail my mind becomes more alert. My perceptions become keener. Each moment I have decisions to make, from where to place my feet so that I do not tumble to how best to navigate the miles ahead. I pay more attention to prominent features of landscape, to sources for water, to potential encounters with animals. It is a wonder, a pleasure. It is also unsettling, for by going off trail I take the risk of getting lost. With similar alertness and feelings, in seeking to understand the human relationship with nature we move off trail into uncharted territory. Questions help set our course. And therein we take our risks.

1

The Biophilia Hypothesis: Empirical Support and Amplifying Evidence

What is biophilia? And why is it important for understanding the human relationship with nature? As for many questions, there is a short answer and a long answer.

The short answer I offered in the introduction: research across many disciplines has been brought together to support the hypothesis that there exists a fundamental, genetically based human need and propensity to affiliate with life. That is biophilia. It is important because if the biophilia hypothesis has merit—and I think it does—it could provide a unifying framework across numerous disciplines to investigate the human relationship with nature. Thus my goal in this chapter—the long answer—is to provide an overview of the literature that makes biophilia compelling. In the next chapter, I approach its shortcomings, and thereby motivate the structural-developmental focus of this book.

Aesthetics and Habitat Selection

By most evolutionary accounts, human beings lived for nearly 2 million years on the savannas of East Africa. During this time, it is believed that certain features of landscape offered greater chances for individual and group survival. For example, bodies of water not only provided a physical necessity to individuals, but also presumably, a perimeter of defense from most natural enemies. Bodies of water also drew forth other animals and plant life on which humans depended. Prominences overlooking grasslands afforded views of approaching threats posed by certain animals or inclement weather. Trees with relatively high canopies did not block the view. Flowers indicated food sources. Based on such an evolutionary account, Wilson (1984) asks rhetorically: "[I]s the mind predisposed to life on the savanna, such that beauty in some fashion can be said to lie in the genes of the beholder?" (p. 109).

Research bears on this proposition. Kaplan and Kaplan, for example, have conducted extensive research on individuals' preferences for different sorts of landscapes (Kaplan and Kaplan 1989; see also, e.g., R. Kaplan 1973, 1977, 1985; and S. Kaplan 1983, 1987, 1992). They found that, in general, people preferred natural environments more than built environments, and built environments with water, trees, and other vegetation more than built environments without such features. Cross-cultural studies by other researchers continue to support this finding (see Ulrich's 1993 review). Not that all natural environments are equally preferred. Kaplan and Kaplan (1989) found, for example, that low action waterscapes were "a highly prized element in the landscape" (p. 9). So were landscapes that were open, yet defined, with "relatively smooth ground texture and trees that help define the depth of the scene" (p. 48). According to Kaplan and Kaplan, such landscapes "can be called parklike or woodlawn or savanna" (p. 48). In contrast, they found that people consistently reported low preferences for settings that were blocked, such as a dense tangle of understory vegetation dominating the foreground of a scene. Such findings did not appear to be directly attributable to a wide variety of competing explanations, such as knowledge about an environment, urban versus rural upbringing, or race.

Early research by Wohlwhill (1968) provided tentative evidence that middle levels of complexity—the richness or number of different objects in the scene—largely explained environmental preferences. Kaplan and Kaplan found partial support for this hypothesis insofar as people did not prefer scenes that lacked complexity. Yet high degrees of complexity did not by itself increase preference. Thus it "is now quite clear that there is more to experimental aesthetics than optimal complexity" (S. Kaplan 1992, p. 595). Kaplan and Kaplan (1989) found, for example, that in judging landscapes people appear "to be heavily influenced by the potential for functioning in the setting. Thus indications of the possibility of entering the setting, of acquiring information, and of maintaining one's orientation emerge as consistently vital attributes" (p. 38). In particular, two important landscape characteristics emerged in their research. One characteristic they call "legibility"—that one could find one's way back if one ventured further into the scene depicted. Such scenes offer visual access, but with distinct and varied objects to provide notable landmarks. A second characteristic they call "mystery"—that one could acquire more information by venturing deeper into the scene and changing one's vantage point. Such scenes include winding paths, meandering streams, and brightly lit areas partially obscured by some foliage.

Kaplan and Kaplan (1989, p. 7) write that these findings can stand apart from evolutionary theory insofar as the psychological affects for preferences for certain landscapes appear real—perhaps "as close to universals as one can find" (p. 150)—and do not require evolutionary theory to be accepted as valid phenomena. Yet in recent years, S. Kaplan has increasingly recognized that the findings are not only consistent with evolutionary theory but also partly explained by it. S. Kaplan (1992) writes, for example, that since trees and water support human survival, "selection pressures in this direction [to prefer trees and water] would hardly be surprising" (p. 587). Moreover, there is "reason to believe that selection pressures in early humans favored acquiring new information about one's environment [mystery] while not straying too far from the known [legibility]" (p. 585).

Along similar lines, Orians and Heerwagen have explicitly drawn on evolutionary theory to frame their research hypotheses on landscape aesthetics. They write: "What we are suggesting . . . is that people have a generalized bias toward savanna-like environments. If this bias does, indeed, exist, then people should react positively to savannas even in the absence of direct experience" (Orians and Heerwagen 1992, p. 560). To investigate this issue, they conducted a cross-cultural study with subjects in the United States, Argentina, and Australia (see Orians and Heerwagen 1992 for a preliminary report). Through a photo questionnaire, they asked three groups of subjects to rate the attractiveness of different kinds of trees. They found that the trees rated as most attractive by all three groups—those with moderately dense canopies and trunks that bifurcated near the ground—matched the prototypic savanna tree. The trees rated as least attractive by all three groups had high trunks and skimpy or very dense canopies. In another study, Orians and Heerwagen (1992) analyzed the changes in landscape that were recommended by a well-known eighteenth-century British landscape architect, Humphrey Repton. They hypothesized that if "humans have an intrinsic bias for particular kinds of landscapes and landscape elements, it should be possible to see this bias in the features and elements that are added to environments to enhance their appeal" (p. 19). Their results supported this hypothesis. Repton, for example, regularly added groups of trees to pastures; and not just any type of tree, but trees that, according to Orians and Heerwagen, resemble the prototypic trees of the savanna.

The savanna hypothesis was tested more directly in a developmental study by Balling and Falk (1982, and summarized by S. Kaplan 1992, and Orians and Heerwagen 1992). Subjects from the eastern United States were asked to rate five different biome stimuli representing rain forest, mixed hardwood forest, boreal forest, East African savanna, and desert. Based on the savanna hypothesis, Balling and Falk hypothesized that the younger children (eight- and eleven-year-olds) would prefer the savanna to the other four biomes. In turn, since familiarity with an environment is a factor in environmental preference, and since increasing age would correspond to children's increasing familiarity with their surrounding hardwoods, it was also hypothesized that the older children would equally prefer the hardwood forests and savanna, and prefer both biomes over the other three. In support of the savanna hypothesis, this pattern of environmental preferences was found.

Taking a somewhat different tack on investigating environmental aesthetics, an unusual study was conducted in a psychiatric hospital in Sweden on the effects of the visual representation of nature. Based on records kept during a fifteen-year period, it was found that patients often complained about many of the paintings and prints displayed in the hospital. Seven times over this fifteen-year period patients attacked a painting or print (e.g., tearing a picture from a wall and smashing the frame). Each time the composition of the painting or print was substantially abstract. In contrast, there was no recorded attack on wall art depicting nature (see Ulrich 1993). These findings are consistent with a study by Ulrich (1993) that found that short-term psychiatric patients responded favorably to wall art that involved nature (a rural landscape or a vase of flowers), but tended to react negatively to abstract painting and prints in which the content was either ambiguous or unintelligible.

Physiological and Psychological Well-Being in Response to Natural Landscapes

If through evolution certain natural landscapes have promoted human survival and reproductive success, then it may have come to pass that such landscapes nurture the human physiology and promote a sense of emotional well-being. Research also bears on this proposition. Findings from over 100 studies, for example, have shown that stress reduction is one of the key perceived benefits of spending time in a wilderness area, especially in those settings that resemble the savanna (Ulrich 1993).

Other studies have examined the relative effects of natural and urban settings in reducing stress. For example, in one study (Ulrich et al. 1991), 120 subjects were exposed to a stressful movie and then to videotapes of either natural or urban settings. Data were collected not only by means of self-report, but through a battery of physiological measures that included heart rate, muscle tension, skin conductance, and pulse transit time. Overall, findings showed greater stress recovery in response to the natural settings.

Other studies conducted in prisons, dental offices, and hospitals point to similar effects. For example, E. O. Moore (1982; cited in Ulrich 1993) found that prison inmates whose cells looked out onto nearby farmlands and forests needed less health care services than inmates whose cells looked out onto the prison yard. In a dental clinic, Heerwagen (1990) presented patients with either a large mural depicting a spatially open natural landscape or no mural at all. Patient data included heart rate measurements and affective self-ratings. Results suggest that patients felt less stressed on days when the mural was present. Others have studied the effects of displaying different ceiling-mounted pictures to presurgical patients who were lying on gurneys (Coss 1990, summarized in Ulrich 1993). Systolic blood pressure was measured in three different conditions: a picture depicting a nature scene that included water; an "exciting" outdoor scene portraying a sailboarder leaning into the wind; and no picture at all. Findings showed that after a relatively brief period of exposure (three to six minutes) the systolic blood pressure levels of presurgical patients were 10–15 points lower in the condition that involved the serene nature picture than in the other two conditions.

Studies conducted in other hospitals have shown related findings. In one fascinating study, Ulrich (1984) examined the potential differences in the recovery of patients after gall bladder surgery depending on whether the patients were assigned to a room with a view of a natural setting (a small stand of deciduous trees) or of a brick wall. Patients were paired on relevant variables that might affect recovery (e.g., age, sex, weight, tobacco use, and previous hospitalization). The results showed that "patients with the natural window view had shorter postoperative hospital stays, and far fewer negative comments in nurses' notes ('patient is upset,' 'needs much encouragement') and tended to have lower scores for minor postsurgical complications such as persistent headache or nausea requiring medication. Moreover, the wall-view patients required many more injections of potent painkillers, whereas the tree-view patients more frequently received weak oral analgesics such as acetaminophen" (Ulrich 1993, p. 107).

In extending this research, Ulrich and Lunden (1990) randomly assigned 166 patients undergoing open-heart surgery with visual stimulation of one of two different types of nature pictures (either an open view with water or a moderately enclosed forest scene), an abstract picture, or a control condition consisting of either a white panel or no picture at all. Their results showed that the patients exposed during surgery to the picture of an open nature view with water experienced much less postoperative anxiety than the control groups and the groups exposed to the other types of pictures. This finding not only points to the effect of natural scenes in promoting recovery but is consistent with the savanna hypothesis, given the lack of recovery when patients were exposed to the enclosed forest picture.

In Kaplan and Kaplan's (1989) reading of the literature of hundreds of studies, they conclude that the "immediate outcomes of contacts with nearby nature include enjoyment, relaxation, and lowered stress levels. In addition, the research results indicate that physical well-being is affected by such contacts. People with access to nearby natural settings have been found to be healthier than other individuals. The longer-term, indirect impacts also include increased levels of satisfaction with one's home, one's job, and with life in general" (p. 173). Later, Kaplan and Kaplan (1989) write that as "psychologists we have heard but little about gardens, about foliage, about forests and farmland. . . . Perhaps this resource for enhancing health, happiness, and wholeness has been neglected long enough" (p. 198). "Viewed as an amenity," Kaplan and Kaplan (1989) write, "nature may be readily replaced by some greater technological achievement. Viewed as an essential bond between humans and other living things, the natural environment has no substitutes" (p. 203).

Affiliation with Animals

If biophilia is understood as an affiliation with the living world, then analyzing our relations to animals should provide data consistent with the biophilia hypothesis. Lawrence (1993) provides one such line of analysis in her lively essay, "The Sacred Bee, the Filthy Pig, and the Bat out of Hell: Animal Symbolism as Cognitive Biophilia." She reasons that if humans have a deeply grounded affiliation with animals, then such affiliation should find pervasive expression in our language and cognition. Even cursory passes at our everyday language point to rich examples. We use "expressions like porker, hogwash, male chauvinist pig, gas hog, road hog, living high on the hog, happy as a pig in muck, going hog wild, piggish, and crying like a stuck pig. There are fascist pigs and Nazi pigs; prostitutes and policemen are called pigs" (p. 325). Claude Lévi-Strauss says that animals are "good to think" as well as good to eat; Lawrence shows that this dictum holds true not only for primitive cultures but for complex modern societies.

As with landscapes, perhaps even more so, it appears that human contact with animals promotes physiological health and emotional well-being. Consider, for example, the common aquariums that—at least in years past—inhabit waiting rooms in many dental offices. Does the conspicuous placement of these aquariums reflect but an arbitrary cultural convention? In a simple experiment, Katcher et al. (1983) found that watching an aquarium resulted in significant decreases in blood pressure below the resting level in both hypertensive and normal subjects. In a more detailed experiment, Katcher, Segal, and Beck (1984) examined the influence of aquarium contemplation on patients about to undergo oral surgery. After the surgery, assessments of the patients' comfort level during surgery were made by the oral surgeon (who was unaware of the nature of the pretreatment), an observer, and the patient. Results showed that aquarium contemplation was as effective as hypnosis in relaxing patients and in increasing their comfort level during surgery. In another study (Beck and Katcher 1996, chap. 1) researchers examined the influence of pets on the course of heart disease. Tracking 92 patients, and accounting for social variables known to be associated with mortality from heart disease, it was found that the mortality rate among people with pets was about one-third that of patients without pets. Contact with animals also positively affects people who have organic or functional mental disorders (see Katcher and Wilkins 1993 for a review). For example, hundreds of clinical reports show that when animals enter the lives of aged patients with chronic brain syndrome (which follows from either Alzheimer's disease or arteriosclerosis), the patients smile and laugh more, and become less hostile to their caretakers and more socially communicative. A number of studies also show that through interactions with animals (such as a dog, cat, bird, dolphin, or even small turtle), autistic children have more focused attention, social interaction, positive emotion, and speech.

In response to this body of research, Shepard (1996) offers grudging acceptance:

Incarcerated incompetents, handicapped outpatients, plain folks who are just getting old, impoverished or stressed executives and their lonely children—all are happier or live longer in the regular presence of friendly animals. There is also less suicide or aggression among the criminally insane, calming among the bereaved, quicker rehabilitation by alcoholics, improved self-esteem among the elderly, increased longevity by cardiac patients and cancer victims, improved emotional states among disturbed children, better morale of the blind or deaf, more cheer among the mentally and physically handicapped, faster learning in the retarded, solace for the terminally ill, and general facilitation of social relationships. (p. 148)

There is a bite in Shepard's writing because, according to Shepard, domestic animals are also, in effect, "biological slaves who cringe and fawn or perform" as we wish (p. 151). Domestic animals "are not a glorious bonus on life; rather they are compensations for something desperately missing," "vestiges and fragments from a time of deep human respect for animals, whose abundance dazzled us in their many renditions of life" (p. 151). Shepard, like Lawrence, believes that wild "animals were among the first objects of classificatory thinking" (p. 97), and that "the human species emerged enacting, dreaming, and thinking [wild] animals and cannot be fully itself without them" (p. 4). Thus, according to Shepard (1993), toward understanding biophilia, the human affiliation with animals needs to be investigated in not only its domestic but its wild forms.

Valuing Nature

Another important line of research that bears on the biophilia hypothesis emerges from Kellert's investigations of people's attitudes and values concerning nature. Through over twenty years of research, Kellert (e.g., 1980, 1983, 1985, 1991, 1993, 1996a, 1997) has refined a typology of nine values, which he suggests "reflect a range of physical, emotional, and intellectual expressions of the biophilic tendency to associate with nature" (Kellert 1996, p. 26).

Although it is difficult to capture briefly the substance of these nine values, definitions provide a starting point. (See Kellert 1996 for his book-length discussion of these values and their place in a larger account of the value of life.) (1) The *utilitarian value* emphasizes the material benefit that humans derive from exploiting nature to satisfy various human needs and desires. (2) The *negativistic value* emphasizes feelings of aversion, fear, and dislike that humans have for nature. (3) The *dominionistic value* emphasizes the desire to subdue and control nature. (4) The *naturalistic value* emphasizes the many satisfactions people obtain from the direct experience of nature and wildlife. (5) The *ecologistic-scientific value* emphasizes the systematic study of the biophysical patterns, structures, and function of nature. (6) The *aesthetic value* emphasizes a primarily emotional response of intense pleasure at the physical beauty of nature. (7) The *symbolic value* emphasizes the tendency for humans to use nature for communication and thought. (8) The *humanistic value* emphasizes the capacity for humans to care for and become intimate with animals. Finally, (9) the *moralistic value* emphasizes the right and wrong conduct toward the nonhuman world.

Based on this typology, Kellert investigated a wide range of differences in values toward nature by age, culture, education, income, ethnicity, gender, and place of residence (urban/rural). Two of these variables are particularly germane to this discussion: age and culture. In terms of age trends, Kellert (1996) says "that children under six years of age were found to be egocentric, domineering, and self-serving in their values of animals and nature, a tendency reflected in especially high utilitarian and dominionistic scores" (p. 47). This young age group also revealed little recognition or appreciation of the autonomous feelings and independence of animals, and expressed the greatest fear of the natural world and indifference toward all but a few familiar creatures. Kellert found that between the ages of six and nine, children became more aware of animals as possessing interests and feelings unrelated to themselves, and realize that animals might suffer pain and distress. Kellert noted the most dramatic increase in children's factual understanding and knowledge of animals and the natural world between the ages of nine and twelve. Finally, between the ages of thirteen and seventeen, children exhibited a sharp increase in ecologistic and moralistic values. For example, adolescents became acutely concerned with conservation and treating other creatures with moral consideration.

In terms of culture, Kellert (1996) considers whether perspectives of the natural world constitute relative expressions of the human condition, or whether there are "only a limited number of ways people can value the living world in a healthy, functional, and sustainable manner" (p. 132). Kellert embarked on a series of cross-cultural studies to investigate this issue. In his research in Japan, for example, Kellert found that in comparison to American populations, the Japanese scored high on a dominionistic value of nature and wildlife. That is, as a whole, the Japanese often sought to manipulate and control nature, and to cultivate preferred natural elements. Thus traditional Japanese "nature appreciation activities—bonsai, haiku, flower arranging, the tea ceremony, rock gardening—reflect a refined appreciation of nature, even at times its veneration, but also a belief that wildness requires the creative hand and eye of humans to achieve its perfection" (Kellert 1996, p. 139). Concordantly, Kellert found that the Japanese people lacked interest in wild nature and ecological processes, and demonstrated limited support for wildlife conservation and protection. In another contrast to the Japanese, Germans demonstrated more pronounced moralistic and ecologistic values, and a greater willingness to subordinate practical needs to maintain pristine nature and protect wildlife. Germans also appeared to romanticize wild nature, stressing its ennobling qualities while having few direct experiences with it, mostly in recreational settings. In short, cross-cultural variability emerged in values of nature and its conservation. Kellert (1996) notes: "This variability appears to be more a matter of degree, however, than any fundamental difference in each nation's basic perspectives of the living world" (p. 145).

Kellert's characterization of these nine values has led to a rich analysis of the human affiliation with nature: biophilia, broadly construed. In recent years, Kellert has also developed the biologically adaptive underpinnings of these values. Insofar as this latter account is successful, it further supports the evolutionary component of the biophilia hypothesis. For example, Kellert suggests that in earlier times many utilitarian activities (e.g., the ability to identify and pick edible berries) provided an unequivocal adaptive advantage by increasing the likelihood of survival (e.g., by increasing nutritional intake). Similarly, the negativistic value presumably helps keep humans a safe distance from dangerous parts of nature such as poisonous snakes and spiders large predators, and precipices. The aesthetic attraction to varying species and landscapes may reflect, as Orians and others have suggested, "a recognition of the increased likelihood of finding food, safety, and security in nature" (Kellert 1996, p. 17). Naturalistic experiences often reduce stress, sharpen sensitivity to detail, enhance creativity, and provide intellectual stimulation and physical fitness. Even the moralistic value, according to Kellert, may provide adaptive advantage by fostering certain forms of kinship, loyalty, and cooperation.

In Kellert's account, while evolutionary biology has an important place, it should not be construed as rigid or deterministic but rather as setting loose parameters in human lives. As Kellert writes: "Although the values of living diversity have been depicted as biological tendencies, experience and culture, as noted, exert a profound influence on their content, direction, and intensity" (p. 37). It is this working of biology, experience, and culture, coupled with his sustained empirical research program, that has led Kellert to eloquent indictments of modern times. "People can survive the extirpation of many life forms," Kellert (1996) writes, "just as they may endure polluted water, fouled air, and contaminated soils. But will this impoverished condition permit people to prosper physically, emotionally, intellectually, and spiritually?" (p. 32). According to Kellert, we destroy nature at our peril, and deceive ourselves in the process: "[N]o society can retain for long its economic or cultural prosperity if it is built upon a despoiled natural world. . . . We must dispel the great fallacy of the modern age that human society no longer requires varied and satisfying connections with the nonhuman world" (pp. 216–17).

"Native" Biophilia in Native Peoples

Many people besides Kellert have argued that the modern world has substantially lost diverse and satisfying connections with nature (to name but a few, Abram 1996; Berry 1977; Leopold 1970; Muir 1976; Mumford 1970; Nabhan and Trimble 1994; Partridge 1984; Roszak 1993; Shepard 1996; Strong 1995; and Thomashow 1995). Based on this assumption, it is reasonable that many have looked to native peoples as a means to understand if not rekindle that connection. This line of reasoning has its parallel in terms of the biophilia hypothesis, and can be set up in the following way: "It's all well and interesting to study biophilia in modern times, but don't you know that many native people never had to study biophilia because they lived it; their lives were deeply connected to nature, their affiliations pervasive across most if not all aspects of their lives. Thus, if you want to study biophilia in any complete sense, study the 'native' biophilia of native peoples."

One of the most sensitive inquiries of this kind has been developed by Nelson (1983, 1989, 1993) in his extensive study of the Koyukon in Northern Alaska. As partial support for the ways in which the Koyukon affiliate with nature, Nelson points to these people's detailed and accurate knowledge of their natural environment. Volumes could be written based entirely on Eskimo knowledge, according to Nelson (1993), about the behavior, ecology, and utilization of arctic animals including polar bear, walrus, bowhead whale, beluga, bearded seal, ringed seal, caribou, musk, and ox. Indeed Nelson suggests that "the expert Inupiaq hunter possesses as much knowledge as a highly trained scientist in our own society, although the information may be of a different sort" (p. 208).

The Koyukon not only have knowledge about animals but appear to learn from them and cooperate with them for mutual advantage. Inuit methods for hunting seals, for example, are essentially identical to those of the polar bear. Nelson (1993) wonders whether this is a case of independent invention, or even convergent evolution, or whether—as I suspect Nelson believes—"Eskimos learned the techniques by watching polar bears, who had perfected an adaptation to the sea ice environment long before humans arrived in the arctic" (p. 210). Or consider another example: The raven is often considered to bring luck if sighted during a hunt, and has been known to lead hunters to their prey. Coincidences? Folklore? Nelson (1989) asks:

Does the raven really care about things, does he really know, does he move with the power Koyukon elders hold in such great regard? . . . [I]f the raven has power, does he recognize it himself and use it consciously? Koyukon hunters say he does. If the raven brings you luck it's to serve himself, because he will eat whatever you leave for him from the kill. (p. 25)

Thus the Koyukon appear to enter with animals into something like a symbiotic relationship.

Such affiliations with animals are woven into the moral and religious fabric of Koyukon life. According to Koyukon elders, no animal should be considered inferior or insignificant. Each deserves respect (Nelson 1989, pp. 23 and 160). Nelson (1993) records that a "Koyukon woman described praying to a raven when she was desperately sick, then explained: 'It's just like talking to God; that's why we pray to ravens'" (p. 214). And Nelson (1989) comes to recognize that when Koyukon speak to the Raven, they draw "on emotions as elemental as devout Christians feel when they pray to God" (pp. 242–243).

According to Koyukon teachers, all are part of a living community. It is a community that includes not only humans, animals, and plants, but mountains, rivers, lakes, storms—the earth itself. Nelson (1989) writes:

According to Koyukon teachers, the tree I lean against feels me, hears what I say about it, and engages me in a moral reciprocity based on responsible use. In their tradition, the forest is both a provider and a community of spiritually empowered beings. There is no emptiness in the forest, no unwatched solitude, no wilderness where a person moves outside moral judgment and law. (p. 13)

Thus, according to Nelson (1993), the Koyukon worldview expands the very meaning of biophilia. It "carries us beyond the idea that humans have a tendency to affiliate with other life . . . to the possibility that our fellow creatures also have a tendency to affiliate with you" (pp. 214–215).

Nelson is trained as a Western anthropologist and is all too aware of the skepticism that greets his research. Yet he persists. He argues that for 99 percent of our history human beings lived exclusively as hunter-gatherers. On a relative time scale, agricultural societies have existed only briefly, urban societies even more briefly.

From this perspective, much of the human lifeway over the past several million years lies beyond the grasp of urbanized Western peoples. And if we hope to understand what is fundamental to that lifeway, we must look to traditions far different from our own . . . Probably no society has been so deeply alienated as ours from the community of nature, has viewed the natural world from a greater distance of mind, has lapsed to a murkier comprehension of its connections with the sustaining environment. Because of this, we are greatly disadvantaged in our efforts to understand the basic human affinity for nonhuman life. Here again, I believe it's essential that we learn from traditional societies, especially those in which most people experience daily and intimate contact with land. . . ." (1993, pp. 202–203)

Conclusion

In this chapter, I have sought to convey what is meant by biophilia, and to make the hypothesis compelling. Thus I reviewed a wide range of literature. No single line of evidence is meant to stand alone. But taken together the research is reasonably impressive, especially since none was conducted under the rubric of the biophilia hypothesis.

The research suggests, for example, that people often prefer natural environments to built environments, and built environments with water, trees, and other vegetation to built environments without such features. These preferences may fit patterns laid down deep in human history on the savannas of East Africa. Experience with or even a visual representation of such landscapes, or key features of landscape, appears to work powerfully within the human physiological and psychological systems. Even minimal experiences with nature can reduce immediate and long-term stress, reduce sickness of prisoners, calm patients before and during surgery, and promote healing after surgery. Direct contact with animals has been shown to greatly benefit a wide range of clinical patients, from adults with Alzheimer's disease to autistic children, as well as children and adults within the general population. Animal images and metaphors appear woven into the fabric of the English language if not the human mind. Moreover, if native peoples offer us a way to understand what is most basic to our being, then the evidence from the Koyukon of Northern Alaska speaks to pervasive affiliations with nature that run deep in our evolutionary history.

No wonder, then, that Wilson (1984) points out that real estate prices are comparatively high for bluff-top land that overlooks bodies of water (lakes, rivers, oceans). If cost is commensurate with desirability, such a finding follows from a biophilic account of aesthetics and habitat selection. Or no wonder, as Wilson notes, that in North America each year more people visit zoos than all sporting events combined. The need and propensity for humans to affiliate with nature appears great.

2

The Biophilia Hypothesis: Conceptual Difficulties and Empirical Limitations

Can biophilia—as its proponents believe—provide a strong framework for an interdisciplinary research agenda on understanding the human relationship with nature?

To answer this question, I examine three concerns at the forefront of current debate. One concern arises if biophilia is understood largely as a genetically determined affiliation. Cast this narrowly, where is the place for cognition, free will, development, and culture? A second concern arises through a seeming contradiction. On first glance biophilia would seem to mean something like "love of nature," or at least something that accentuates a positive affiliation. Yet most proponents of biophilia agree that people at times find nature unlikable and unfriendly, if not threatening and harmful, and that such negative affiliations comprise a part of biophilia. How can these ideas be reconciled? A third concern arises when biophilia is understood to be vigorously shaped by experience, learning, and culture. Cast this broadly, can the biophilia hypothesis ever be disconfirmed? How good is the supporting evidence?

Biophilia and Genetic Determinism

One of the fundamental concerns voiced about the biophilia hypothesis centers on the extent to which biophilia is genetically shaped, if not determined. In a review, for example, in *Science* of Kellert and Wilson's (1993) edited book *The Biophilia Hypothesis*, Fischer (1994) opens by saying that "Wilson and his colleagues have identified yet another human behavior they suspect is governed by genes—nature appreciation" (p. 1161). According to Fischer, to the extent the biophilia hypothesis embodies a genetic determinism, the hypothesis is deeply flawed; to the extent it does not, the hypothesis is "largely hollow" (p. 1161). Most critiques of this sort follow on the heels of a more general critique of sociobiology. Thus it will be useful if we first attend to this broader literature.

The sociobiological program is often said to have been launched with the publication of Wilson's (1975) *Sociobiology: The New Synthesis*. In this book, Wilson argues that our complex behavioral responses are little more than genetically programmed behaviors to maximize genetic fitness. This perspective can be well illustrated by sketching the answer to what Wilson and others call "the problem of altruism": If it is true that people act so as to increase their reproductive fitness, how can altruistic behavior be explained, because it seemingly reflects behavior that decreases an individual's reproductive fitness for the sake of others. Sociobiologists have suggested two types of answers: reciprocal altruism and kin selection. Roughly stated, Trivers (1971) has proposed that often when we help others we do so because at some point we ourselves may be in need of help. In other words, because of reciprocal altruism, a certain amount of helping behavior is in fact in our best interest. Other times we also help genetically related family members because they share common genes, and to help them is to increase the reproductive success of our common gene pool. "This enhancement of kin-network welfare in the midst of a population is called kin selection" (Wilson 1975, p. 116; compare Hamilton 1964).

Accordingly, Wilson (1975) says that it may well be "that the time has come for ethics to be removed temporarily from the hands of the philosophers and biologized" (p. 287). Elsewhere Wilson says that "[m]orality, or more strictly our belief in morality, is merely an adaptation put in place to further our reproductive ends. . . . In an important sense, ethics as we understand it is an illusion fobbed off on us by our genes to get us to cooperate" (Ruse and Wilson, 1985, pp. 51–52). In other words, we have been genetically programmed to think and feel as if we freely make ethical choices; but such thoughts and feelings are epiphenomenal in that they offer no authentic causal explanation for our behavior.

Moving beyond ethics, Wilson boldly concludes his 1975 treatise by saying that in the final analysis psychology, sociology, and the other human sciences will be reducible to neurobiological processes: "Cognition will be translated into circuitry. Learning and creativeness will be defined as the alteration of specific portions of the cognitive machinery. . . . To maintain the species indefinitely we are compelled to drive toward total knowledge, right down to the levels of the neuron and gene" (p. 575). According to Wilson, when "we have progressed enough to explain ourselves in these mechanistic terms," the result—"a world divested of illusions"—might be hard to accept, but true (p. 575).

Wilson anticipated rightly. Such a result has been hard for many people to accept. Over the last two decades, at least two approaches toward critiquing sociobiology have emerged.

Through one approach, critics have gone after the sociobiological program directly and taken it to task for substantive flaws in its methods, science, logic, and use of evidence. Consider a common method that sociobiologists employ to build their theory. Some pattern of behavior in animal species is found that appears similar to what people do. Then both pieces of behavior are named the same thing. Then evidence is acquired for the existence in animals of a genetic predisposition to the behavior, and then sociobiologists announce the general result that there is genetic basis for this type of behavior in humans (see Kitcher 1985, p. 185).

In the literature, sociobiologists have used "rape," for example, to describe not only human behavior but the behavior of scorpions, flies, and ducks. Barash (1979) says that among ducks, "sometimes strange males surprise a mated female and attempt to force an immediate copulation, without engaging in any of the normal courtship ritual and despite her obvious and vigorous protest. If that's not rape, it is certainly very much like it" (quoted in Kitcher 1985, p. 185). Barash then suggests that through "rape" male ducks might maximize their genetic fitness; and that perhaps "human rapists, in their own criminally misguided way, are similarly doing the best they can to maximize their fitness" (quoted in Kitcher 1985, p. 186).

The problem is that an enormous and often unsubstantiated jump occurs in moving from animals to humans. In a counterargument, Kitcher (1985) suggests that rape has complicated social and not just genetic origins that involve prevalent social conceptions of the role of female and male status. Applying the term "rape" to certain instances of copulation among ducks is to misunderstand the distinctly human origins of rape and to trivialize such explanations. Moreover, even on genetic grounds Kitcher calls into question whether male hominids would increase their genetic fitness through rape. After all, rapists are frequently attacked or punished, and accordingly there should be a decrease in their reproductive success. In addition, rapes frequently take place on humans who cannot conceive (e.g., children, women past the age of menopause, and members of the same sex). "Of course," Kitcher writes, "sociobiologists could contend that such behavior is a by-product of mechanisms that were selected under different conditions" (p. 187). But Kitcher remains unconvinced in that such "special pleading takes cover in our ignorance of the hominid social environment, in an effort to accommodate troublesome facts that are all too obvious" (p. 187). Many others have similarly been concerned with the sociobiological bent to tell "just so" stories—post hoc accounts without enough specificity to rule out competing and seemingly more compelling explanations (Diamond 1993; Katcher and Wilkins 1993; Fischer 1994).

While the above approach seeks to critique the sociobiological program directly, another approach seeks to reinstate the primacy of nonbiological constructs. For example, imagine three situations where a fellow named Bob goes for a walk in the woods. In the first situation, a boulder above Bob is dislodged by an earth tremor; it falls on Bob and kills him. As unfortunate as this situation is for Bob, we would not normally say the boulder acted immorally. In the second situation, a lion on the hunt pounces from out of the woods and kills Bob (and feeds her family on him). Again, from Bob's standpoint this situation is most unfortunate; but we would not normally say that the lion acted immorally, even though the lion is a biological animal. Finally, in the third situation, a bandit jumps out of the woods and kills Bob in the process of robbing him. It is only in this latter situation that we normally would say something unethical or immoral has occurred.

Why do we say this? Presumably because we believe that ethics is something distinctly human and that an account of our ethical judgment extends beyond our biological base as an animal. Granted, there may be no way to prove such an account, just as there may be no way to prove that you, the reader, are not actually at this moment a brain in a vat being stimulated by electrical currents. But the phenomenological experience of your life speaks against the latter position, as the phenomenological experience of our ethical judgments speaks against the reduction of ethics to biology. Thus in critiquing sociobiology, this other approach seeks to reinstate the primacy of such central human constructs as altruism, morality, free will, and human agency—to reinstate them in the sense that such constructs play authentic causal roles in human lives.

Concerns like these just begin to touch on the enormous quantity of critical literature on sociobiology. It is this literature that will increasingly bear down, justifiably so, on the biophilia hypothesis, if the hypothesis is conceived in stringent biological terms. Yet it is not clear that Wilson or his colleagues do so. In an unpublished response to Fischer's (1994) critique of biophilia noted above, Kellert (1994) argues strongly that Fischer has "oversimplified and exaggerated the biological determinism articulated" in the book.

Kellert's point is well taken, for Wilson at times seems to recognize the paramount importance of learning, cognition, and culture in explaining human nature. For example, Wilson (1993) says that biophilic behavior, "like other patterns of complex behavior, is likely to be mediated by rules of prepared and counterprepared learning" and that the multiple strands of the biophilic emotional response "are woven into symbols composing a large part of culture" (p. 31). Furthermore, consider the diverse range of contributions that Kellert and Wilson solicited for their edited volume on the biophilia hypothesis. Many of the contributions focused on the religion and practices of native peoples (Diamond 1993; Gadgil 1993; Nelson 1993; Nabhan and St. Antoine 1993), myths (Shepard 1993), language symbolism (Lawrence 1993), cognitive processing (Ulrich 1993), and environmental philosophy (Rolston 1993). Hardly a reductionist collection. Moreover, Wilson writes—often eloquently—of seemingly irreducible qualities of human life; that is, qualities largely unleashed from their biological components. For example, Wilson (1992) writes that an "enduring environmental ethic will aim to preserve not only the health and freedom of our species, but access to the world in which the human spirit was born" (p. 351). Before that he writes that we "do not understand ourselves yet and descend farther from heaven's air if we forget how much the natural world means to us. Signals abound that the loss of life's diversity endangers not just the body but the spirit" (p. 351). Elsewhere, Wilson (1984) writes that "our spirit is woven from it [biophilia], hope rises on its currents" (p. 1). Freedom, spirit, heaven's air, and hope rises on biophilic currents. These are not human constructs easily understood in terms of a genetic biological determinism.

At the same time, what are we to make of other parts of Wilson's even recent writings? In one of his books Wilson (1993) asks how biophilia could have evolved. He answers that the likely answer is biocultural evolution, during which culture was elaborated

under the influence of hereditary learning propensities while the genes prescribing the propensities were spread by natural selection in a cultural context . . . a certain genotype makes a behavioral response more likely, the response enhances survival and reproductive fitness, the genotype consequently spreads through the population, and the behavioral response grows more frequent. (pp. 32–33)

This explanation embodies the hard biological stance of the early socio-biological program: Genes that lead to behaviors that enhance survival tend to reproduce themselves (since they are in bodies that procreate more rather than less), and thus these genes and correlative behaviors grow more frequent. In this sense, human behavior (including biophilic behavior) is orchestrated, if not directed and determined, by genes.

Thus in different places Wilson offers differing accounts of biophilia and its genetic basis. But the more important issue is this: If biophilia is understood in stringent biological terms, it will increasingly run up against the formidable critiques that have been extensively leveled against the sociobiological program, especially in its early forms. And there is no need for biophilia to be understood in this way. Rather there are softer biological accounts of biophilia, and they need not be "largely hollow."

Biophilia and Biophobia: A Contradiction?

Poisonous snakes and spiders often frighten people. So do grizzly bears, mountain lions, and other large predators. Mosquitoes are nuisances at best. Bogs and swamps can seem unappealing, precipices scary, dark woods forbidding. As S. Kaplan (1995) says in his review of Kellert and Wilson's (1993) edited volume on biophilia, while "there does appear to be evidence for special feelings" for nature, "some are negative rather than positive" (p. 801). What then becomes of biophilia that, as a word, connotes only the positive? Do negative affiliations with nature contradict the biophilia hypothesis? Such questions lead S. Kaplan to write that it "is hard to shake the feeling that more wishful thinking is going on [with the biophilia hypothesis] than is healthy in competent scholars confronting a difficult problem" (p. 801).

One response has been to separate biophilia (a positive affiliation with nature) from what becomes called "biophobia" (a negative affiliation with nature). Orr (1993), for example, says that biophobia "ranges from discomfort in 'natural' places to active scorn for whatever is not man-made, managed, or air-conditioned. . . . [It is the] urge to affiliate with technology, human artifacts, and solely with human interests regarding the natural world" (p. 416). Elsewhere Orr says that in the same way that love stands in contrast to hate, and life to death, so must we choose biophilia over biophobia. Biophobia "is not OK because it is the foundation for a politics of domination and exploitation" (p. 420). Thus, by this separation, Orr is able to speak of both positive and negative affiliations with nature, and value each differently.

Ulrich (1993) similarly contrasts biophilia and biophobia, but he does so to try to provide a clean empirical method for investigating both constructs. Ulrich, for example, reports on numerous conditioning and counter-conditioning experiments using biophobic stimuli (e.g., pictures of snakes, spiders, heights, closed spaces, and blood). The research suggests that "humans are biologically prepared to acquire and especially to not 'forget' adaptive biophobic (fear/avoidance) responses to certain natural stimuli and situations. . . . Moreover, recent findings suggest that processing of biologically prepared fear-relevant natural stimuli can be very fast and may often occur automatically or 'unconsciously'" (p. 85). In his article, Ulrich provides a remarkable synthesis of hundreds of studies on negative affiliations with certain aspects of nature (biophobia) and contrasts those studies with emerging research (summarized earlier) on positive affiliations with nature (biophilia).

Interestingly, however, Wilson himself sets up a different relation between biophilia and biophobia. Recall that Wilson defines biophilia as an *affiliation* with life and lifelike processes. In this framing of biophilia, people have both positive and negative affiliations with life. These complex feelings "fall along several emotional spectra: from attraction to aversion, from awe to indifference, from peacefulness to fear-driven anxiety" (Wilson 1993, p. 31). Thus, according to Wilson, biophilia says something about the sense of integration of life-affirming and life-disaffirming propensities.

If we take seriously that biophilia subsumes both life-affirming and life-disaffirming propensities, a puzzle arises. How can we know which actions are part of the resulting conservation ethic? Consider, for example, a vivid anecdote that Diamond (1993) provides of his experiences with native peoples in New Guinea:

> I found men intentionally inflicting pain on captured live bats for no other reason than amusement at the reactions of the tortured animals. The men had tied twenty-six small *Syconycteris* blossom bats to strings. They lowered one bat after another until it touched the red-hot embers of a fire, causing the bat to writhe and squeal in pain. The men raised the bat, lowered it again for another touch to the red-hot embers, repeated this process until it was dead, and then went on to the next bat, finding the whole proceedings funny. (pp. 263–264)

Can one torture animals and say that through disaffirming that animal's life one is biophilic? Why not? Presumably it is this sort of problem that led Orr to separate biophilia from biophobia, and argue for the ethical superiority of the former. Or consider a situation that might appear justifiable on evolutionary grounds: logging the virgin timber in the Amazon rain forests. After all, as Ulrich (1993) notes, the Amazon forests are associated with greater risks than the African savannas because of the rain forest's "higher levels of biophobic properties, including spatial enclosure and higher probabilities of encountering close hidden threats, including snakes, spiders, and other fear-relevant stimuli" (p. 118). Thus it would "naturally" follow that humans would want to clear-cut the dense forests and recreate them into savannalike settings. Does such a biological account justify the current deforestation of the Amazon?

I assume Wilson has answers, and likely enough here is one of them: Since certain life-affirming actions (perhaps feeding one's dog) and life-disaffirming actions (perhaps shooting a rabid animal) can promote our genetic fitness, whenever we seek to determine which actions to pursue, all we need to determine is which actions do so. Swatting mosquitoes may promote our genetic fitness, torturing bats may not; sustainable logging might, clear-cutting rain forests may not; and so on. This is what I take Wilson (1984) to mean when he says that the "only way to make a conservation ethic work is to ground it in ultimately selfish reasoning—but the premises must be of a new and more potent kind" (p. 131). To this point, the sophistication of Wilson's theorizing illuminates in the sense that we are better positioned to understand and investigate the complex psychological coordinations of seemingly disparate reactions to nature ("from attraction to aversion, from awe to indifference, from peacefulness to fear-driven anxiety").

But Wilson's account also reduces biophilia to the selfish gene and thus becomes weakened by various forms of the counterarguments to the early sociobiological program sketched in the previous section. Namely, a problem arises in trying to use biophilia as a biological proposition to justify biophilia as a normative or moral proposition. To do so is to commit the naturalistic fallacy (cf. Hume [1751] 1983; Moore, G. E., [1903] 1978): to reason that an account of what "is" necessarily leads to an "ought," that natural forms necessarily make the moral. To underscore this problem, consider Wilson's position on aggression. According to Wilson (1978), "primitive warfare evolved by selective retention of traits that increase the inclusive genetic fitness of human beings" (p. 112). At some point in human warfare, aggression is counterproductive, according to Wilson; that is, the energy expended and the risk of injury and death outweigh the energy saved and the increase in survival and reproduction. But until that point—at which aggression no longer confers genetic advantage—a good deal of blood-filled battles can be fought, women raped, children killed, with "moral" sanction. The word moral is in quotation marks because Wilson, in effect, suggests that if an action confers genetic advantage it is moral, which appears to trivialize the term. Thus I agree with Myers (1996, 1998) that if biophilia is to have moral standing, it needs a theoretical foundation that extends beyond its genetic base.

Mediated Biophilia

If biophilia involved only genetic explanations, then the conceptual difficulties and disconfirming evidence would be, in my estimation, insurmountable. Of course, Wilson (usually), Kellert, and others would agree.

As noted earlier, Kellert counters the critic who goes after a hard-wired, deterministic conception of biophilia by responding that that conception is a straw man, and not at all what is meant by biophilia. Fair enough. But what happens with biophilia in this biologically weaker form—what I will refer to as "mediated" biophilia—where, in Kellert's (1996) words, "experience and culture . . . exert a profound influence on [the] content, direction, and intensity" of "biological tendencies" (p. 37)? How well does mediated biophilia withstand scrutiny? I consider six critiques.

1. Biophilic Symbolism. One line of evidence for the biophilia hypothesis reviewed earlier is that animals (and more broadly, nature) find pervasive expression in our language and cognition. Indeed some proponents of biophilia suggest that the human need for metaphorical expression finds its greatest fulfillment

> through reference to the animal kingdom. No other realm affords such vivid expression of symbolic concepts. The more vehement their feelings, the more surely do people articulate them in animal terms, demonstrating the strong propensity that may be described as cognitive biophilia. Indeed, it is remarkable to contemplate the paucity of other categories for conceptual frames of reference, so preeminent, widespread, and enduring is the habit of symbolizing in terms of animals. (Lawrence 1993, p. 301)

In other words, not only is it proposed that animals (and nature) have an important role in human language and cognition, but that "no other realm affords such vivid expression of symbolic concepts." Is Lawrence correct?

Consider but one counterexample developed by Lakoff (1987) that involves concepts of lust. This example seems particularly relevant because according to adaptive evolutionary theorists, humans should and do think a lot about sex (Bjorklund and Kipp 1996; Buss 1992; Ellis 1992; Wilson 1978). If Lawrence is correct, concepts of lust should find expression through animal imagery. And they do. Lakoff (1987, pp. 409–411, in collaboration here with Zoltan Kovecses) notes some of the following expressions: "Don't touch me, you animal!" "He's a wolf." "Stop pawing me!" "Hello, my little chickadee." "You bring out the beast in me." "He preys upon unsuspecting women." "Wanna nuzzle up close?" "She's a tigress in bed."

But as Lakoff demonstrates, the concept of lust also finds rich expression through nonanimal metaphors. *Lust is heat:* "I've got the hots for her." "She's an old flame." "Hey, baby, light my fire." "She's frigid." "She's hot stuff." "Don't be cold to me." "I'm burning with desire." *Lust is insanity:* "I'm crazy about her." "I'm madly in love with him." "You're driving me insane." "He's a real sex maniac." "She's sex-crazed." *Lust is a game:* "I think I'm going to score tonight." "I struck out last night." "She wouldn't play ball." *Lust is a machine:* "You turn me on." "I got my motor runnin', baby." "Don't leave me idling." *Lust is a force:* "I was knocked off my feet." "When she grows up, she'll be a knockout." "She bowled me over." "She sparked my interest." "I could feel the electricity between us."

This game (if I can use that metaphor) of trading metaphors to speak for or against the biophilia hypothesis could go on for some time. Critics might more productively offer an alternative theory that subsumes animal symbolism, as Lakoff does. In Lakoff's (1987) account, thought "is embodied, that is, the structures used to put together our conceptual systems grow out of bodily experience and make sense in terms of it; moreover, the core of our conceptual systems is directly grounded in perception, body movement, and experience of a physical and social character" (p. xiv). Thus it would follow in Lakoff's view that human symbolism involves animals because they, along with so many other things—body parts, colors, emotions, and the enormous range of human artifacts, for example—are part of the human experience. If Lakoff's account of categorization is even roughly correct, it calls into question Lawrence's (1993) assertion that no other realm besides animals "affords such vivid expression of symbolic concepts" (p. 301). Accordingly, such support for mediated biophilia is weakened.

A Biophilia Rejoinder.   The response here seems relatively simple, and in my view correct: Lawrence just goes too far. That is, I assume a great deal more evidence than I have reviewed here—indeed overwhelming evidence—could be marshaled to show that animals (and nature) have an important place in human language and cognition. That is the key proposition. This proposition does not preclude the importance of other categories. Indeed, in this framing of the biophilia hypothesis, Lakoff's work could potentially be used to help ground what Lawrence calls "cognitive biophilia." Granted, Lakoff subsumes concepts of animals within a larger framework that posits structuring categories of the mind. But since such cognition according to Lakoff is embodied—literally, as in a biological organism—it would appear only a modest stretch to evolutionary theory in general, and the biophilia hypothesis in particular.

2. The Attraction of the "Unnatural." Another critique comes on the heels of the previous one and can be set up in the following way: Even if we grant that people have an affinity for nature, it would seem equally clear that people have an affinity for human artifacts (Rothenberg 1993). People enjoy looking at cityscapes from a distance: of Manhattan, Paris, and San Francisco. People enjoy visiting specific buildings: the Eiffel Tower or Lincoln Memorial. People can spend lavish amounts of money buying cars, and lavish amounts of time polishing and looking at them. People go to great efforts to learn musical instruments and to perform for appreciative audiences. People paint, sculpt, and write novels. People spend hours each day watching television, or on the World Wide Web, garnering information, computer chatting, playing games.

If people have an affiliation with nature and nonnature, then the very construct of biophilia becomes increasingly difficult to understand in a meaningful way. More so, because according to Wilson, biophilia can include negative affiliations: aversions to snakes, spiders, precipices, stagnant water, dark caves. Similarly, it could be said that humans have aversions to standing in front of moving cars, guns, and certain other human artifacts. Thus, the critic might say, the biophilia hypothesis reduces to the following proposition: People affiliate positively, negatively, or neutrally with things natural and human made, outside their own bodies, and including their own bodies. In turn, the critic might argue, such a proposition becomes almost a tautology, and says virtually nothing meaningful about the human condition.

A Biophilia Rejoinder. Consider the proposition that humans affiliate with members of the opposite sex. This proposition seems patently true. Yet presumably there is a tremendous amount to say about such affiliations. Presumably a man can love a woman, or hate her, or maybe even both at the same time, and vice versa. Maybe the way men affiliate with women is somewhat different than the way women affiliate with men. Presumably such affiliations have a tremendous range of expressions (trust, intimacy, passion, flirtation, tenderness) in a vast array of different relationships (between new friends, old friends, lovers, in-laws, coworkers). Maybe such affiliations have an evolutionary basis. Maybe such affiliations are mediated by culture, context, and experience. My point is that there are a lot of important questions that involve understanding the nature of the human affiliation between members of the opposite sex. Those questions are not trivialized by showing the range or complexity of the affiliations, or by showing that such affiliations occur alongside (or are integrated with) other sorts of affiliations (e.g., that a woman might choose to live with a man and a cat). Rather, the affiliations need to be characterized and explained.

The same with biophilia. To say that people affiliate with nature is not to say that people only affiliate with nature. This rejoinder parallels the previous one. Moreover, it could be said that people often find human-made artifacts more appealing when the artifacts are portrayed, encompassed, or used in a natural setting. For example, although it is true that some people can spend lavish amounts of money buying cars, think of the diverse range of natural settings employed by car advertisers (Armstrong 1996): the Jeep Wrangler alone on a bluff top in the southwest deserts of the United States; the Cadillac Seville passing through pastoral farmland—a horse lopes up to the car; the Toyota Tacoma on the beach, on the edge of sand and water. Again and again, advertisers use nature to sell human artifacts. Such an observation does not trivialize the human affiliation with nature, but speaks for it, however it is to be understood. Accordingly, proponents of biophilia will say that what is needed are rich, textured characterizations of, and then explanations for, the human affiliation with nature, and then to integrate those accounts into a wider analysis of culture and human development.

3. The "Unattraction" of the Natural. In the preceding critique, one grants that people at times affiliate with nature. But one need not grant that initial premise. Namely, it could be said that some people actively dislike nature. For example, in characterizing Woody Allen, Mia Farrow once observed:

Woody has no tolerance for the country. . . . Within half an hour after arriving he's walked around the lake and is ready to go home [to New York City]. He gets very bored. . . . Of course, he never goes in the lake, he wouldn't touch the lake. "There are live things in there," he says. (Partridge 1996, p. 158)

In turn, Partridge (1996) calls Woody Allen a "puzzling counterexample" to the biophilia hypothesis.

A Biophilia Rejoinder. Two questions need to be asked here. First, how many people actually exist who, like Woody Allen, appear actively to dislike nature? Quite possibly the answer is very few. If so, such counter-examples would not refute the biophilia hypothesis: no more than do the existence of religious hermitages refute the general statement that adults tend to engage in procreative relationships. Second, to what extent do people who purport to dislike nature actually dislike nature? Granted, we can easily imagine that people raised exclusively in urban environments might feel uncomfortable if thrown into a wilderness setting. But would such people prefer living on an urban street lined with trees or not? Do they enjoy flowers in their home? Do they have a pet dog or cat? Do they ever visit parks, zoos, or aquariums? Do they find sunsets pleasant to watch? If affirmative answers begin to emerge from some collection of such questions, then their dislike for nature is not nearly as pervasive as initially claimed. Moreover, even if negative answers emerge, what one has assessed are individuals' subjective perspectives on their relationships with nature. In turn, it is possible (given the research literature surveyed in chapter 1) that by lacking substantive contact with nature, such individuals unknowingly suffer for it in terms of their overall physical and psychological health.

4. The Savanna Hypothesis. The critic might agree with at least some of the research reviewed earlier that provides tentative support for the savanna hypothesis—that our affiliations with landscapes fit patterns laid down deep in human history on the savannas of East Africa. The same critic, however, might also point to what appears to be a great deal of disconfirming evidence. What of people who travel great distances to spend time not on savanna-like bluff tops but on tropical beaches? In snow-covered alpine resorts? In the rain forests of Costa Rica or Brazil? People spend sizable amounts of money to sail on Alaskan cruise ships, and appear to enjoy watching glaciers calf amid the land of the midnight sun. Other people build houses in the redwood forests of the Pacific Northwest. Again and again, people appear to have affinities for diverse natural landscapes far removed from the African savannas. Such observations seem to disconfirm the savanna hypothesis.

A Biophilia Rejoinder.   There are a handful of possible responses. One response is to suggest that affiliations with nonsavanna landscapes often mimic key features of the savanna. For example, beaches provide wide open landscapes (visibility), a barrier for protection, and water; and alpine areas provide the elevation, relief, and (often) visibility associated with the savanna. Another response is to suggest that certain regions are attractive for reasons largely independent of the landscape itself. A biologist might take delight in an upcoming trip to the Amazon, with visions of discovering hundreds of new insect species. A cultural anthropologist (or traveler of similar ilk) might look forward to refreshingly different cultural experiences. But how many people actually consider a trip to the Amazon landscape rejuvenating—as a desirable place, for example, to retire? Very few. Another response is to make the case that for people living for many thousands of years in an unsavannalike setting (such as the Amazon River basin), genetic changes occur such that aspects of those landscapes that promote survival benefits become aesthetically pleasing and rejuvenating. Diamond (1993) develops this response when he says: "I am puzzled, in other discussions of the biophilia hypothesis, by what seems to me an exaggerated focus on savanna habitats as a postulated influence on innate human responses. Humans spread out of Africa's savannas at least 1 million years ago. We have had plenty of time since then—tens of thousands of generations—to replace any original innate responses to savanna with innate responses to the new habitats encountered" (pp. 253–254). In other words, the evolutionary account can hold, but the savanna hypothesis needs to give way to a broader account of genetic predispositions to inhabited landscapes. A final approach—if and when the evolutionary biology comes up short—is to recognize the place of culture, experience, and learning, all of which can profoundly mediate the attraction to savannalike landscapes.

5. Can the Biophilia Hypothesis Be Disconfirmed?   The above possible counterexplanations regarding the savanna hypothesis (if not all of the above critiques) highlight a particularly troubling aspect of the biophilia hypothesis: the seeming inability to provide disconfirming evidence. When an initial biophilic hypothesis is not confirmed, a post hoc explanation is offered. Each explanation may well be reasonable, but there is an uncomfortable proliferation of them; moreover, when the going gets rough, proponents of biophilia can just wave the wand of culture, experience, and learning.

It is a familiar problem, one that critics have articulated powerfully and repeatedly in seeking to discredit Freudian theory (see, e.g., Blight 1981; Crews 1995; Grünbaum 1984). The charge is that psychoanalytic theory can accommodate virtually any set of findings. A conscious desire for a boy to kill his father is evidence for the oedipal complex. If such a desire is not recognized, that becomes evidence for the constructs of repression and the unconscious. Ditto, the critic may charge, with biophilia. The biophilia hypothesis is slippery and difficult to understand except in metaphorical (unscientific, nontestable) terms.

A Biophilia Rejoinder.   Notice the language Kellert and Wilson (1993) use in titling their edited volume. They call it *The Biophilia Hypothesis,* not something like *The Biophilia Revolution.* Why? Because they want to encourage scientific investigations of biophilia across the natural and social sciences. So, yes, I assume they would agree that the biophilia hypothesis should move forward by means of both framing testable hypotheses and critiquing the existing relevant literature for, at times, insufficient methods.

Yet I think this scientific orientation will only go so far. To illustrate this limitation, recall one experiment reviewed earlier. Katcher et al. (1984) found that watching an aquarium resulted in significant decreases in blood pressure below the resting level in both hypertensive and normal subjects. Is this really evidence for biophilia? Where are the control conditions? Let us imagine for a moment that with adequate controls it was found that not just fish in an aquarium but "slow-moving globs of multicolored light" decreased blood pressure. Would this disconfirm the biophilia hypothesis? No, at a minimum the finding would simply not support the biophilia hypothesis. After all, many activities—such as listening to a Mozart sonata or taking an afternoon nap—presumably lower blood pressure. The biophilia hypothesis does not preclude such effects, no more than it precludes—as discussed earlier—human artifacts from playing an important role in human cognition and symbolism. It is equally possible in our imaginary experiment that slow-moving globs of multicolored light decrease blood pressure because such blobs mimic natural states (fish in water). In this case nature is primary, the artifact secondary, and one more item would be added to the biophilic repertoire of evidence. Even if experiments could distinguish between these two interpretations—and I am not convinced this is an empirical issue—it would still not satisfy the empirical scientist. For as long as a nature effect is found in enough relevant situations, biophilia cannot lose. In other words, no single experiment of this sort provides disconfirming evidence; at best it can provide only nonconfirming evidence.

6. Evidence that Tugs at the Evolutionary Component of Biophilia.

Even in mediated biophilia, evolutionary theory has an important—albeit reduced—place in an account of the human affiliation with nature. Yet several lines of research tug even at that reduced place. For instance, from decades of ornithological research in New Guinea, Diamond (1993) suggests that while the New Guineans have a profound knowledge of nature, they exhibit virtually no positive emotional responses to it: no "love, reverence, fondness, concern, or sympathy" (p. 262). Earlier, I quoted Diamond's account of the New Guineans who found enjoyment in torturing bats. Diamond also reports on the young New Guineans who left behind the Stone Age technology of their parents to find work in urban areas. According to Diamond, these young people exhibit little interest in the surrounding national parks and zoos, "negligible interest in the natural heritage with which their ancestors lived so intimately for tens of thousands of years" (p. 269), and fear of the forest. Where, then, is the strength of the biology if an interest for nature can be so easily lost in a single generation—nay, even within a few years?

These parts of Diamond's essay have attracted the attention of critics of the biophilia hypothesis (Fischer 1994). Yet while Diamond's credentials as an ornithologist are undisputed, he appears to have little training as an anthropologist or cross-cultural psychologist insofar as he reports only anecdotal social-scientific data. Thus some may find Diamond's conclusions suspect, although suggestive.

That said, research by Nabhan and St. Antoine (1993, see also Nabhan and Trimble 1994, chap. 5) provide supporting evidence. Nabhan and St. Antoine investigated responses to the natural world across two groups of Uto Aztecan cultures along the U.S./Mexico desert borderlands: the O'odham and the Yaqui. Within each culture they compared responses between "tribal elders, who have engaged in considerable hunting and gathering activities during their lifetimes, with those of their grandchildren who have grown up fully exposed to television, prepackaged foods, and other trappings of modern life" (p. 230). By holding constant the genetic lineages, their study provides a means to assess the place of cultural and environmental influences on the expression of biophilia.

Based on a variety of measures, Nabhan and St. Antoine found that the O'odham and Yaqui children's knowledge of, interest in, and appreciation for their natural world was strikingly at odds with that of their grandparents. For example, despite access to open spaces, the majority of the O'odham and Yaqui children had never spent more than half an hour alone in a wild place. Television provided children with more exposure to wild animals than did their natural surroundings. And large percentages of children did not know basic facts of desert life: that it is possible, for example, to eat the fruit of the prickly pear cactus—a major food source in their lands for more than 8,000 years.

In terms of the genetic component of the biophilia hypothesis, what is striking is that within one or two generations seemingly deep and pervasive affiliations of the O'odham and Yaqui with nature have been considerably extinguished. Perhaps Diamond is correct after all; perhaps the genetic basis of biophilia is smaller than initially proposed by Wilson, and more needs to be said about development and culture.

A Biophilia Rejoinder.   Those who are committed to a genetic account of biophilia will seek an alternative explanation. In interpreting their findings, for example, Nabhan and St. Antoine suggest that the genetic predisposition for biophilia exists, but its expression needs to be triggered by the culture and environment. Specifically, they propose that the O'odham and Yaqui have been losing three requisite triggering mechanisms: the loss of biodiversity in the deserts; the loss of hands-on, visceral contact with nature; and the loss of the oral traditions of plants and animal stories. Nabhan and St. Antoine may be correct, but a lot of work is being asked of the undeveloped notion of "triggering" (compare Campbell and Bickhard 1987). Usually a trigger involves something simple that sets into motion complicated, extensive, and/or powerful happenings. But Nabhan and St. Antoine seem to offer the substance on the cultural and environmental level, and the simplicity on the level of the gene. More troubling, their specific data, as far as I can tell, provides no direct evidence for a genetic influence. In other words, a theorist working biophilia exclusively on the cultural level could look at Nabhan and Antoine's results and say, "Yes, that's exactly what I would have expected."

Conclusion

Where do we now stand in an assessment of the biophilia hypothesis? In my interpretation the research literature speaks relatively strongly for the proposition that people have a need and propensity to affiliate with nature. That is biophilia, in its least developed and least controversial form. It also is clear that humans affiliate both positively and negatively with nature. It is not unreasonable to call the former affiliations "biophilia" and the latter affiliations "biophobia." However, I think in the long run that approach will not be as productive, let alone elegant, as following the lead of Wilson, Kellert, and others, letting biophilia refer to both positive and negative affiliations, and then taking up the task of integrating both within a larger framework. In building that framework, I have argued that biophilia needs a theoretical foundation that extends well beyond evolutionary biology (see Kahn 1997a). In turn, that is what we can now move to: theory and research on the human relationship with nature, focusing on development and culture.

3

The Psychological Framework: Structure and Development

In the opening to Plato's (1956) *Meno,* a young Thessalian nobleman, Menon, asks a question: "Can you tell me, Socrates—can virtue be taught? Or if not, does it come by practice? Or does it come neither by practice nor by teaching, but do people get it by nature, or in some other way?" (p. 28). In response, Socrates says:

My good man, you must think I am inspired! Virtue? Can it be taught? Or how does it come? Do I know that? So far from knowing whether it can be taught or can't be taught, I don't know even the least little thing about virtue, I don't even know what virtue is! . . . [T]ell me yourself, in heaven's name, Menon, what do you say virtue is? (p. 29)

Thus, as he is wont to do, Socrates sets into motion a dialogue. Moreover, Socrates analytically distinguishes the psychological question (how virtue develops) from the philosophical question (what virtue is), and argues that the latter needs as much attention as possible before addressing the former.

I like Socrates' overall distinction, and shall but reverse the order to serve the purposes of this book. Namely, in the next chapter I address the questions of what virtue—or more broadly morality—is, and why moral theory is important for understanding the human relationship with nature. Here I present the structural-developmental framework that grounds my research. Because some readers may not have expertise in psychology, I begin by offering some orienting words about developmental theory in general. I then explain why, in particular, one might have intellectual commitments toward structural-developmental theory. Finally, I take up what structural-developmental theory means by the idea of psychological structures and hierarchical integration, and what the theory has to say about the relation between judgment and action.

An Overview of Developmental Theories

Broadly speaking, at least four overarching types of explanations can be provided for how people develop: endogenous, exogenous, causal inter-actional, and structural developmental (see Kohlberg and Mayer 1972, J. Langer 1969, and Turiel 1983, for further explications of this typology). Endogenous theories propose that development largely occurs through internal mechanisms. In the previous chapters we examined one such theory in the context of the biophilia hypothesis: Sociobiologists maintain that we behave in ways that promote our survival, and that such behavior has been genetically programmed through an evolutionary process of selection. Other endogenous theories take a more humanistic turn. For example, Rousseau (1964) suggests "that man is naturally good" (p. 23). A. S. Neill built his school Summerhill based on a similar view. Neill ([1960] 1977) says: "My view is that a child is innately wise and realistic. If left to himself without adult suggestion of any kind, he will develop as far as he is capable of developing" (p. 4). Notice that while sociobiology and the largely undeveloped nativistic views of Rousseau and Neill feel very different (at a minimum, Rousseau and Neill posit an innate goodness or wisdom in a way that sociobiology does not), they share a view that the mechanism for development lies within each individual.

In contrast, exogenous theories propose that development largely occurs through external mechanisms. Included here are behavioristic theories that focus on stimulus-response mechanisms and operant conditioning (Skinner [1948] 1976, 1971, 1974), social-learning theories that focus on modeling and imitation (Bandura 1977; Bandura and Walters 1963; Rushton 1982; Staub 1971), and transmission theories of moral character (Bennett and Delatree 1978; Ryan 1989; Walberg and Wynne 1989; Wynne 1979, 1986; Wynne and Ryan 1993). For example, in establishing behaviorism as field of study, Watson ([1924] 1970) says: "Give me a dozen healthy infants, well-formed, and my own specified world to bring them up in and I'll guarantee to take any one at random and train him to become any type of specialist I might select—doctor, lawyer, artist, merchant-chief and, yes, even beggar-man and thief, regardless of his talents, penchants, tendencies, abilities, vocations, and race of ancestors" (p. 104). Watson here oversteps most behaviorists who believe in certain biological constraints. But Watson's point remains well taken insofar as behaviorists focus on the external contingencies of learning. This view is of a piece with other exogenous theorists who conceive of the child as largely passive, reminiscent of John Locke's tabula rasa—a blank slate upon whom adults directly transmit socially correct knowledge and moral precepts.

The distinction so far can be thought of in terms of nature or nurture. Endogenous is nature. Exogenous is nurture. These are two of the choices that Menon offered Socrates. In addition, there are hybrids of both: that is, theories that try to explain the causes of human behavior in terms of endogenous and exogenous factors. Freud ([1920] 1967, [1923] 1960, [1924] 1963, [1930] 1961) offers a clear example of such a theory. Consider Freud's answer to Meno's question. According to Freud, at around five years of age a boy wants to sleep with his mother (and for this issue Freud mostly focused on boys, not girls). This desire derives much of its strength from the sexual instinct, but prevented from finding expression by environmental conditions, especially by the father. Thus, according to Freud, toward the father the boy feels aggression, which also has a strong instinctual base. Stated forcefully, the boy wants to kill his father, setting up what Freud refers to as the oedipal complex. But the boy not only hates his father, but loves his father, and these two emotions together intensify the conflict. Toward resolving this conflict, the boy identifies with the father and internalizes the father's standards and morals. Through this internalization, a new psychic entity develops: the superego—the boy's moral conscience. Or, stated as an answer to Meno's question, virtue develops, according to Freud, through this interaction between endogenous and exogenous factors.

Finally, there are structural-developmental theories, also sometimes referred to as constructivist, social cognitive, or structural-interactional. Structural-developmental theories posit that development is grounded in human knowledge and values, in the active mental life of children as they construct increasingly more adequate ways of understanding their world and of acting upon it. Structural-developmental theories appear similar to causal interactional theories (such as offered by Freud) insofar as they both substantively account for endogenous and exogenous factors. But one important distinction is that, at least as proposed by Piaget, structural-developmental theory seeks not to understand causes of behavior, but to use behavior as the means to characterize the developing structure of the human mind. I will say more about this distinction shortly, in terms of how structural-developmental theory contends with potential conflicts between one's judgment and corresponding action.

Structural-Developmental Theory

But first, allow me to convey for the uninitiated Piagetian a better sense of the structural-developmental project. I start by sketching one of Piaget's classic experiments on conservation. Then I discuss in more depth the idea of psychological structures and of hierarchical integration. As will become apparent in subsequent chapters, I employ both ideas vigorously in my research on the human relationship with nature.

In conducting Piaget's conservation experiment, a child (let us say of five years) is shown two identical beakers (A and B) filled with the same amount of water. The child is asked whether there is more water in beaker A, more water in beaker B, or the same amount of water in both beakers. In virtually all cases the child answers: the same amount in both beakers. Next, in front of the child, all of the water from beaker B is poured into a third beaker (C) that is taller and narrower. The empty beaker B is then removed from the child's sight. Thus in front of the child now rests beakers A and C. The experimenter now asks a pivotal question: Is there more water in beaker A, more water in beaker C, or the same amount of water in both beakers? The nonconserving child (often under six years of age) says that the two beakers do not have the same amount of water, and typically that there is more water in beaker C because it is taller. But perhaps the child only confuses the word "more" with "taller." So Piaget ([1952] 1965) probes. He asks, for example, "Is there more to drink, or does it just look as if there is?" The child answers: "There is more to drink" (p. 7). Piaget offers pages of similar dialogue. His claim is that the nonconserving child considers "the quantity of liquid to vary according to the form and dimensions of the containers into which it is poured.

Perception of the apparent changes is therefore not corrected by a system of relations that ensures invariance of quantity" (p. 5). In turn, the conserving child establishes the invariance of quantity by various means, including identity (it is the same water), reversibility (pouring the water from beaker C back to beaker B would reestablish the initial condition), and compensation (beaker C is taller, but it is also narrower).

Can one teach a child the concept of conservation? Here is a story told to me, supposedly true. Two professors of psychology were friends. One professor had commitments to Piagetian theory, the other to behavioristic theory. The behaviorist had looked at the Piagetian literature and thought to himself,"Hmmm, you don't need Piagetian theory to explain these stages of development; rather, children's understandings and correlative behaviors can simply be taught." He thereby devised a plan. He had two children at home, a five-year-old son and a three-year-old daughter. He decided to teach conservation to his son, and demonstrate the results to his friend, the Piagetian scholar.

So, one evening after dinner the behaviorist scholar asked his son to play a little game with him. The father then set up beakers of water and replicated the methodology (above) that Piaget employed in helping to assess conservation. At the pivotal point in the "game" he asked his son, "Is there more water in beaker A, more water in beaker C, or the same amount of water in both?" As expected, his son provided a nonconserving response. The father then calmly explained to his son that he (the son) had the wrong answer, and explained the right answer, and provided appropriate reasons. His son then repeated the right answers and appropriate reasons. The father then repeated the game over the course of several evenings, and changed the size of the beakers so as to promote the transference of his son's knowledge to new situations. The father thought: "I did it; I taught conservation to my son, and it wasn't really hard at all."

The behaviorist scholar then invited the Piagetian scholar over for a relaxing dinner. After dinner, the behaviorist scholar told his friend, "Oh, by the way, I have a little something to show you." He then called his children into the room, and set up the conservation task. At the pivotal point in the task, he again asked his son, "Is there more water in beaker A, more water in beaker C, or the same amount of water in both?" In response, his son said there was the same amount of water in both beakers, and then provided a reason based on compensation: although beaker C was taller, it was also thinner. The father then broke out into a big smile. But, as this story goes, as the father was beaming, his son turned to his little sister and whispered (in a voice all could hear): "But they don't really have the same amount of water, you know."

What happened? One interpretation is that the son learned what DeVries (1988) calls "school varnish"—facts largely divorced from cognition. More recently, DeVries (1997) writes: "When the content of societal values (rules) and truths is not understood, a child can only assimilate it to the schemes he or she has constructed and can only approximate the observable form but not the substance of an adult's proposition. The result is that the child's thought may not be really transformed but changed only superficially" (p. 10). In turn, it is the above sort of anecdotal evidence that can lead one to constructivist psychological commitments—to a view that knowledge (or learning) is not simply the accumulation of facts, or conditioned responses, or conditioned behavior. Rather, while knowledge entails facts and behavior, it more fundamentally entails understanding, and children actively construct their understandings through interaction with the physical and social world.

Why do children construct knowledge? Consider an older infant who has developed a grasping scheme that works well for her. She sees a small ball, reaches with one hand, and picks it up. This behavior works time and again. Indeed, she can pick up a small ball with either hand. But then let us say that one day she encounters a ball that is too big to pick up with one hand. Her grasping scheme does not work—her knowledge of how to pick up balls is not adequate to solve this problem. That is disconcerting. And it is disconcerting not only to a parent who wants his daughter to succeed, or to an educator who wants to teach this infant, but most important to the infant herself. She is disequilibrated. She wants to pick up that ball. The interest and desire is there. Thus she struggles for a more adequate understanding. She experiments. Maybe she tries the other hand, and that fails. At some point she discovers a solution. She uses two hands. She takes two different grasping schemes and coordinates them into a single consolidated scheme, and thereby picks up the larger ball with both hands. Disequilibration is followed by the construction of knowledge.

According to Piaget ([1952] 1963), an infant is born with innate reflexes of an extremely plastic nature. These reflexes can be thought of as biological schemes—biological patterns of action. Through the functions of assimilation and accommodation, these biological schemes make interaction possible and thus lead to the construction of psychological schemes—general organized psychological systems of action. Though Piaget says it is difficult to determine when exactly biological schemes can be differentiated from psychological schemes, he suggests this criterion: biological schemes are associated with the undifferentiation of assimilation and accommodation, while psychological schemes with their differentiation. For example, when an infant first encounters his or her mother's breast, it is not clear whether the sucking reflex assimilates the breast or accommodates to it. Thus the sucking is still biological. A short time later, however, the infant can be observed to seek actively for the mother's nipple, to differentiate it from other parts of the mother's body, and overall to assimilate the mother's nipple into what can now be called an early psychological sucking scheme. In this way, initial biological reflexes are transformed into organized psychological patterns of thought. In turn, from this point on in development it is primarily the psychological concepts themselves that undergo further transformations and consolidation.

Psychological Structures

The idea of psychological structures can be partly conveyed by means of an analogy to physical structures, like houses. First, we can note that houses do not exist spontaneously in nature (and here I am excluding such potential domiciles as caves and hollowed-out tree trunks). Rather, we construct houses. (Indeed we call the industry of making houses "the construction industry.") Second, our constructions aim to serve certain functional goals: We want to keep out the rain, and build a roof. We want to keep out the wind, and build walls. We want to let in light, and build windows. We want to get in and out easily, and build doors. Third, these functional goals (and subsequent constructions) emerge through our interactions with the environment. For example, in a cold, sunny climate we might seek to build a house that lets in sunshine, while in a hot, humid climate we might seek to keep sunshine out. Fourth, based on such functional interactions, we can characterize the structural similarities of the houses of diverse people. An igloo, an adobe house, and a Frank Lloyd Wright masterpiece all share certain things: roofs, insulation, doors, interior demarcations for living spaces, and so forth. We can also characterize contextual differences insofar as such roofs, insulation, doors, and interior spaces often look different due to different types of interactions people have with nature, and the building materials available. After all, it is not easy to build an igloo in Texas. Similarly, the structural-developmental proposition is that we construct our psychological structures through interaction with our environment; the environment can be seen to constrain and prod but not cause development.

More formally, Piaget (1970) says that a structure can be understood in terms of "three key ideas: the idea of wholeness, the idea of transformation, and the idea of self-regulation" (p. 5). Wholeness refers to a general system of thought that is applied to a wide variety of tasks, transformation to the changing of that system, and self-regulation to the processes (entailing assimilation and accommodation) by which such transformations occur. Broadly speaking, these properties and their corresponding structures can be found in Piaget's account of the four overarching stages of cognitive development—sensorimotor, preoperations, concrete operations, and formal operations—where through development there is an increasing capacity for synthesis and abstraction. Although this is not the place to characterize each of these stages (see, e.g., Cowan 1978; Flavell 1963; Ginsburg and Opper 1969; Lourenço and Machado 1996; Piaget 1983), it is important to discuss briefly what Piaget refers to as the hallmark of concrete operations: reversibility. I will harken back to it in subsequent chapters.

Reversibility refers to ideas that, as Chapman (1988) writes, "are linked together such that in the passage from one state to another, the first state is not lost but retained as a simultaneous possibility that could potentially be reinstated" (p. 45). A clear example of reversibility is revealed by one of Piaget's class inclusion tasks. Show a preoperational child a necklace comprised of 20 brown wooden beads and 7 white wooden beads. Ask the child to count the brown beads. Twenty. Ask the child to count the white beads. Seven. Ask the child to count all the wooden beads. Twenty-seven. Ask the child whether there are more brown beads or more white beads. More brown beads. Now ask the child this pivotal question: "Are there more wooden beads or more brown beads?" The preoperational child will answer that there are more brown beads. Why? According to Piaget ([1952] 1965), it is because the child cannot simultaneously hold in his or her mind both the class of wooden beads and the class of brown beads. Thus, when the child decomposes the class of wooden beads into the class of brown beads and white beads, the child can only compare the class of brown beads to what remains: the class of white beads. Only by concrete operations, with the onset of reversibility, can the child, in effect, simultaneously decompose and re-compose the initial class of wooden beads.

This example from concrete operations also helps to provide more of a feel for the developmental aspect of Piaget's theory. For example, at the preoperational level, the dog Spot and the dog Buffy might make up the class of dogs in a child's house. So instead of $A + A' = B$, we have $a + a' = c$, where a and a' are elements instead of classes, and c but a subset of A. Developmental continuity can be seen insofar as the elements that make up class A are what comprise the objects of thought in preopera-tions. Developmental discontinuity can be seen in the qualitative differ-ences in the composition of the groups. In this way, there is not so much a right answer to what counts as the "real" structure of thought, but rather that, once specified, different levels of analyses provide charac-terizations of corresponding levels of organization (compare Flavell 1963, chap. 5). This approach reflects the meaning behind structural-developmental research, where structures themselves are conceived of not as static organizations, but as dynamically situated within an organism's intelligence.

Hierarchical Integration

The two terms "hierarchical" and "integration" each have their place in the types of biological thought discussed in chapters 1 and 2. Drawing on Simon (1969), Pinker and Bloom (1992), for example, suggest that "hierarchical organization characterizes many neural systems, perhaps any system, living or nonliving, that we would want to call complex"

(p. 485). Integration of perspectives is exactly what Cosmides, Tooby, and Barkow (1992) argue for across the biological and social scientific disciplines. But in structural-developmental theory the two ideas come together to comprise the idea of a structural transformation where through development later structures reflect increasingly adequate transformations of earlier structures. I touched on this idea in the above example in which the elements of classificatory reasoning in concrete operations were the objects of conceptualization in preoperations.

I now extend this idea of hierarchical integration from the logical domain into the social and moral domains. To do so, I make use of Kohlberg's work in moral development (e.g., Kohlberg 1969, 1971, 1980, 1984, 1985). As the reader may know, building on Piaget ([1932] 1969) Kohlberg interviewed children, adolescents, and adults using hypothetical moral dilemmas. Based on his research, Kohlberg proposed that moral reasoning followed a developmental pathway composed of six hierarchically integrated stages. In stage 1 Kohlberg proposed that there is moral consideration only for the self (punishment avoidance). Stage 2 involves the selfishness of stage 1, but works it within a broader framework where one can be instrumentally concerned about others (instrumental hedonism), as in, "I'll scratch your back if you scratch mine." Stage 3 incorporates the relational exchange aspects of stage 2, but grounds that in a genuine care and concern for the interpersonal relationship, particularly in the context of family members and friends. As stage 3 can be seen to correct for limitations of stage 2 (a tit-for-tat morality), so does stage 4 correct for the limitations of stage 3. After all, what of a judge who presides over a case that involves family members or friends, and who gives them preferential treatment? Is that fair? What of our moral obligations to strangers whom we meet (the homeless mother on the street) or do not meet (political prisoners tortured in China)? Stage 4 morality addresses these concerns by reworking considerations of self, others (instrumentally), and family and friends by codifying the demands of justice through largely impersonal rules and laws within organizational systems. Yet the limitation here is that such rigid systems fail to articulate underlying moral principles that also allow for their flexibility and judicious application. Thus there is the movement toward stages 5 and 6, toward the articulation of a generalizable theory of human rights and justice. It is in this sense—in which the central organizing principle of an earlier stage becomes an element in a more inclusive organizational framework—that stages represent hierarchical transformations of moral knowledge, rather than simply replacements of one moral view with another.

I am well aware that Kohlberg's theory, like that of many other pivotal theorists in a field, has been critiqued often and hard. It has been argued, for example, that Kohlberg's theory fails to take adequate account of regression (Selman, Jaquette, and Lavin 1977), is not as universal as claimed (Shweder, Mahapatra, and Miller 1987; Snarey 1985), confounds conventions with morality (Turiel 1983), and provides an inadequate characterization of moral caring (Gilligan 1982; Noddings 1984), moral character (Campbell and Christopher 1996), and prosocial development (Eisenberg-Berg 1979; Eisenberg and Fabes 1998). Most of these critiques have some merit; a few of the critiques a great deal. I shall return to some of these critiques in chapter 4. But without worrying at this point about the validity of Kohlberg's actual stages, Kohlberg's orientation to studying moral development through structural-developmental analysis remains highly influential today (see, e.g., Helwig 1997; Killen and Hart 1995; Kurtines and Gewirtz 1991; and Puka 1995); and Kohlberg's approach, at least in my judgment, withstands the torrent of over three decades of criticism.

Moreover, I think Kohlberg's use of the idea of hierarchical integration offers ways to reconceive old problems and solve new ones. For an old problem, recall one of the questions that sociobiological theory has sought to answer: If it is true, as sociobiologists maintain, that people act so as to increase their reproductive fitness, then how does sociobiological theory explain altruistic behavior, since such behavior would seemingly decrease an individual's reproductive fitness? As described in chapter 2, sociobiologists have suggested two types of answers: reciprocal altruism and kin selection. Both answers attempt to show that some degree of helping others is actually in the actor's genetic self-interest.

That is one answer to "the problem of altruism," but it is not the only one. Consider a situation where a man spends every weekend helping his elderly and cranky aunt, and in the process forgoes his pleasurable weekend activities. His actions appear altruistic. But let us say that the aunt is quite wealthy, and that the person says that the only reason he helps his aunt is because he hopes to inherit millions of dollars when the old lady finally dies. Given such a reason, the act now appears selfish. Which is it? Drawing on Baldwin ([1897] 1973), from a psychological standpoint such a situation is framed wrongly. For according to Baldwin, egoism and altruism are fundamentally linked: "The ego and the alter are thus born together. Both are crude and unreflective, largely organic. And the two get purified and clarified together. . . . My sense of myself grows by imitation of you, and my sense of yourself grows in terms of my sense of myself. Both ego and alter are thus essentially social" (p. 9). That is, from the start a child is neither entirely selfish nor entirely altruistic—both go together, because both depend on one another in the child's development. Thus instead of following sociobiologists and trying to reduce altruism to selfishness ("the selfish gene"), it is possible to account for both, but in a hierarchically integrated sense.

Here is how. Wilson (1975) says that the idea of reciprocal altruism is "expressed in the familiar utterances of everyday life," which take the form: "Come to my aid this time, and I'll be your friend when you need one" (p. 553). Like Wilson, Kohlberg would agree that such reasoning is essentially selfish. But based on scores of developmental studies, Kohlberg would also say that such reasoning does not reflect a developmental endpoint. Rather, as one's interests expand to include not only one's own self (stage 1), another's self instrumentally (stage 2), one's family and friends (stage 3), one's society (stage 4), and the world (stages 5 and 6), the "self" is not lost but embedded in a larger "alter." Indeed, in specific circumstances, the self can still assert itself in "individualistic" ways. For example, when being sexually assaulted, a person may well lash out with fury and fight. Such "self-oriented" action need not be interpreted as a "regression"—a backward movement along a developmental pathway—but the assertion of self in a context-specific situation. Thus, in terms of hierarchical integration, selfishness does not disappear but gets transformed into a more complex, more adequate, and qualitatively different form of ethical understanding.

That is a sense of what hierarchical integration offers for an old problem. Now for a new one. Recall one of the difficult problems for the biophilia hypothesis: how to understand the relationship between individuals' negative and positive affiliations with nature. The research shows (and common sense confirms) that people can fear and dislike aspects of nature, particularly those that are threatening to the self, physically and psychologically. People can also enjoy controlling and subduing nature. Both this negativistic and dominionistic orientation—using Kellert's (1996) typology—need not disappear through the lifespan. But if "biophilia development" occurs, it is possible that these orientations become hierarchically integrated into a more comprehensive orientation. Along these lines, consider Nelson's (1989) description of his experience after killing a deer in his homeland in Alaska:

I whisper thanks to the animal, hoping I might be worthy of it, worthy of carrying on the life it has given, worthy of sharing in the larger life of which the deer and I are a part. Incompatible emotions clash inside me—elation and remorse, excitement and sorrow, gratitude and shame. It's always this way: the sudden encounter with death, the shock that overrides the cushioning of the intellect. I force away the sadness and remember that death is the spark that keeps life itself aflame. (p. 263)

Nelson notes his feelings of "elation" and "excitement" in the kill, in the domination over nature, except it is no longer domination but an affiliation with nature that weaves his own life into that which he has killed: "worthy of sharing in the larger life of which the deer and I are a part."

On a philosophical level, Rolston (1989) works similar ideas. He says:

[W]e must judiciously blend what I call "natural resistance" and "natural conductance." Part of nature opposes life, increases entropy, kills, rots, destroys. Human life, like all other life, must struggle against its environment, and I much admire the human conquest of nature. However, I take this dominion to be something to which we are naturally impelled and for which we are naturally well-equipped. Furthermore, this struggle can be resorbed into a natural conductance, for nature has both generated us and provided us with life support—and she has stimulated us into culture by her resistance. Nature is not all ferocity and indifference. She is also the bosom out of which we have come, and she remains our life partner, a realm of otherness for where we have the deepest need. I resist nature, and readily for my purposes amend and repair it. I fight disease and death, cold and hunger—and yet somehow come to feel that wildness is not only, not finally, the pressing night. Rather wildness with me and in me kindles fires against the night. . . . Environmental life, including human life, is nursed in struggle; and to me it is increasingly inconceivable that it could, or should, be otherwise. If nature is good, it must be both an assisting and a resisting reality. We cannot succeed unless it can defeat us. (pp. 49–50)

Rolston keenly recognizes that parts of nature are to be feared (as it "opposes life," "kills," and "destroys"), and parts of nature are to be dominated (an activity for "which we are naturally impelled" and "naturally well-equipped"). But in Rolston's account, both are hierarchi-cally integrated in a life-affirming orientation.

I am not saying that the above account characterizes a developmental aspect of biophilia—although it might. That proposition would need the support of developmental research. But I am trying to quicken interest in the idea of hierarchical integration, and to show how it can be employed in developmental analyses.

The Relation between Judgment and Action

As noted earlier in this chapter, some theorists—usually of an exogenous orientation—have proposed that behavior is the central construct of importance. Many decades ago, for example, Watson ([1924] 1970) said, "Behaviorism . . . holds that the subject matter of human psychology is the behavior of the human being. Behaviorism claims that consciousness is neither a definite nor a usable concept" (p. 2). Similarly, in the field of education, character educational theorists have argued for programs that emphasize inculcating socially acceptable behavior (Ryan 1989; Walberg and Wynne 1989; Wynne and Ryan 1993). What does it matter, they ask, whether a student believes it is immoral to cheat on a test, or immoral to pick a fight with a younger student? The important question is, does the student cheat and fight, and if so, how can educators and parents change that behavior?

Behavior is important, but a theory of behavior divorced from reason-ing can only come up short. In his landmark article critiquing the behav-iorist project, Chomsky (1959) writes that one "would naturally expect that prediction of the behavior of a complex organism (or machine) would require, in addition to information about external stimulation, knowledge of the internal structure of the organism, the ways in which it processes input information and organizes its own behavior" (p. 27). This proposition is clearly true for a machine. When I sit down at my computer, I can, by external stimulation (tapping various keys on my keyboard), get the computer to perform many tasks: word processing, statistical analysis, e-mail, and so forth. But I know virtually nothing about computer programming, about how the software works. And I know virtually nothing about the hardware. When the computer mal-functions, can I fix it? Usually, no, because I have little understanding of the internal workings of the machine. Similarly—and this is Chomsky's point—to understand human beings we need to understand how people process information and structure their own behavior.

As I mentioned earlier, Piaget's agenda was not directly tied to understanding the causes of behavior. But, that said, it is difficult not to ask causal questions of structural-developmental theory, especially when it comes to moral issues. Why did this child hit his friend? Why did that child tell a lie? Why did that adolescent join a gang that preys on the elderly? Why did this other adolescent volunteer at an AIDS hospice? These are sensible questions, especially to parents and educators. Even if structural-developmental theorists did not set out initially to answer such questions, we can still reasonably ask what the theory has to say about them.

The structural-developmental response begins by reframing the question (compare Blasi 1980; Turiel 1983; Turiel, in preparation). Instead of asking whether, and if so how, judgments cause behavior, the question becomes, What is the relation between judgment and action? (Here I use the word "action" instead of "behavior" to help distance the question from the commonly assumed causal epistemology of exogenous theories.) One answer is that often there is a very close relation. I make a judgment that I am thirsty, and then drink a glass of water. I believe it will rain in the afternoon, and bring along my umbrella. I reason that getting an education will further my personal growth and career opportunities, and then enroll at a university. I think it is wrong to rob a bank, and I have never robbed a bank. In these and thousands of other situations there would seem to be a perfect correlation between my judgments and actions.

Framed this way, it is the incongruities that stand out and call for an explanation. How can a child say it is wrong to steal, but steal anyway? How can a husband say he loves his wife, and yet subject her to physical or emotional abuse? How can people reason that it is wrong to pollute the air, but still drive their cars? Part of the answer lies in understanding that people have multiple judgments, and at times these judgments can conflict with one another.

This point was demonstrated elegantly in Asch's (1952) landmark experiments where he investigated ways in which groups can modify and distort individuals' judgments. In one experiment, a subject is placed in a room with 7–9 other persons for the purported reason of participating in a perceptual experimental task. The subject thinks the other people are subjects just like him, though they are actually confederates of the experimenter. The group is told that they will see a series of figures. In each figure they are to compare the length of the standard line to the length of three different comparison lines. In each case they are to say which comparison line is equal in height to the standard line. The task has been constructed so that it is perceptually obvious which one of the comparison lines matches the length of the standard line. The experimenter tells them that since the group is small, each person can just verbally announce his or her answer. The experimenter then goes in an order such that the subject is nearly the last person to give an answer. The experiment proceeds with two trials in which the group (and subject) gives the right answer. Then, in what is referred to as the critical trial, the group gives a wrong answer. They choose a line that is obviously different in length from the standard line, but they assert it is the same length. The experimental question is whether the subject gives a factually correct answer (contrary to the group) or a factually incorrect answer (in accord with the group).

In the critical trials, Asch found that the subjects would go along with the group judgment about one-third of the time. On the individual level, about 12% of the subjects (subsequently termed the "yielding" group) went along with the group all or virtually all of the time, while 42% of the subjects (subsequently termed the "independent group") did not go along with any of the group answers.

Based on his interviews with the subjects after the experiment, Asch found that one of two types of orientations characterized subjects from the yielding group. Some subjects believed that their perception was faulty and that the group had to be correct. Other subjects became relatively unconcerned about the correctness or not of their answer, and were primarily concerned with not appearing different from the group. One subject, for example, said, "I did not want to be apart from the group; I did not want to look like a fool. . . . Scientifically speaking, I was acting improperly, but my feeling of not wanting to contradict the group overcame me" (p. 472). In turn, Asch found some common ground across both independent and yielding groups. No subject disregarded the group judgment; that is, all subjects recognized that the group diverged from their own judgment, and that this posed a difficulty or conflict that needed to be addressed. Moreover, all of the subjects looked for some obvious sources of misunderstanding, such as whether they understood the directions correctly, or had themselves taken a close enough look at the lines.

Taking these results together with the results from Asch's other experimental conditions (e.g., increasing the magnitude of the discrepancy of the comparison line, and introducing a partner with the subject), Asch proposes that social forces do not directly determine an individual's action. Rather, according to Asch, individuals are in interaction with their social environment, and in the context of that interaction individuals make complex judgments about group action and the relation between the group and individual. As Asch says: "Social life makes a double demand upon us: to rely upon others with trust and to become individuals who can assert our own reality" (p. 499). In other words, in asking what the relation is between a person's judgment and action in the Asch experiment, we cannot just look at the relation between an individual's cognitive judgment and his or her action (whether they yield or not to the group judgment). For Asch showed that these individuals not only had cognitive judgments about line height, but also other types of social judgments, including those that pertained to peer acceptance, embarrassment, asserting the truth, and how much truth value rests with majority opinion.

Conclusion

I have sought to provide a sense of the psychological framework that I will draw upon for the remainder of this book. In my thinking, a good deal of this framework is just common sense. People construct conceptual knowledge. Those constructions are mentally organized. We can call those mental organizations structures. Structures develop. Through structural development early forms of knowledge do not disappear, but are transformed into more comprehensive and adequate ways of understanding the world, and of acting upon it. I suggested that this latter idea—of hierarchical integration—is a powerful one, and can offer ways to reconceive old problems. This idea can also help provide answers to new problems, as I will seek to do in later chapters. In addition, I examined the issue of the relation between judgment and action. The important point here is that people have diverse and sometimes conflicting judgments, and that an analysis of human action cannot stand apart from them. Accordingly, in the next chapter I provide some specificity about how to demarcate different forms of social and moral judgments. Then I take up this project more fully throughout the empirical chapters, wherein I seek detailed characterizations of the development of diverse and sometimes conflicting environmental judgments.

4
Obligatory and Discretionary Morality

It's about 15 degrees outside in Cambridge, Massachusetts. Zoe (my then two-year-old daughter) and I frequent the cafes. And we walk. We see a man. He sits on the sidewalk, wrapped in blankets. Zoe and I go up to him. I put a dollar in his hat, and he looks up and smiles. So, we sit down with him. He explains that he's in the middle of a story that he's writing, and that perhaps we would like to hear it, if we would. . . . He's hoping. Sure. It's a story about a prince in rags who needs money for armor and a helmet. He reads with enthusiasm, and with his various gestures it's clear that the allegory is intentional. It's cold. And windy. The sun shines and it's still cold and windy. What about tonight? What can we say? Have a nice day, sir. I say something but I don't remember what. We leave. Fifty yards later the conversation starts:

"Papa, the man, he no have home?"
"No, Zoe, the man doesn't have a home."
"Why not, papa?"
"He just doesn't, sweetie."
"No bath?"
"No, the man doesn't have a hot bath to go to either."
"Why not?"
"Well, he doesn't have any money."
"Money? Why no money?"
"He just doesn't. But we helped him a little bit. We gave him a little money."
"No bath?"

It is a common encounter in many large cities. Homeless people request money from people passing by. Like many others during such encounters, I ask myself questions: Should I give the person some money? Would he use the money to buy food, or instead perhaps alcohol or drugs? Would he hurt me if I approach him? Should not society at large offer more help? Regardless, am I myself morally obligated here and now to give this person some money? What if he is a she, and what if she has a young child with her? If I give money to this person, then how about to the next homeless person or family down the block? Across town? Across continents?

In this chapter, I examine several moral concepts that underlie many of these questions. My reason is that moral issues are often environmental issues, and vice versa. Homelessness, for example, is a moral issue; but it is made so partly because of the type of interaction homeless people have with the natural world. It is cold. And windy. And the man in the above vignette might freeze. Or, casting the environmental problem first, unsustainable farming practices have led to the famine of entire populations. More generally, in countless cases, locally and globally, harm to the natural environment leads to harmful and sometimes unfair consequences to humans. Thus to investigate the human relationship with nature, one needs to bring forward a moral perspective.

Morality, of course, is a contentious topic. Recall from chapter 2, for example, that sociobiologists believe that morality is an illusion fobbed off on us by our genes to get us to cooperate. Others disagree. Philosophers and lay people alike often hold strong and diverging opinions about morality's content, scope, and epistemic status.

Still, the possibilities of what counts as moral usually fall within two broad orientations. One moral orientation focuses on obligatory requirements of right action, and is embodied in most moral theories today (including those that are consequentialist and deontological). The second moral orientation, with roots to Aristotle in *Nichomachean Ethics*, focuses on what it means to be a "good" person or on a conception of the "good" or the "good life." Here the focus is on long-term character traits—such as courage, temperance, and wisdom—and personality. Both orientations, I believe, are central to the moral life of children and adults. For the most part, however, developmental psychologists—like the moral philosophers—have investigated morality in terms of the first orientation. Thus, based on the idea of discretionary morality (where an act is morally worthy but not required), I will explain how I have extended the boundaries of moral developmental research to include concerns of virtue and the "good." In subsequent chapters, I will then use this distinction between obligatory and discretionary morality to help guide my environmental psychological studies.

Obligatory Morality

Moral obligation has been and continues to be fundamental to moral philosophical thought. Around 400 B.C., for example, the Chinese philosopher Mo Tzu established a system of thought that "ranked with Confucianism for some two centuries as one of the eminent schools of the day" (Mei 1972, p. 410). Mo Tzu critiqued Chinese society, including its feudalistic hierarchy and daily brutalities, and offered a solution. He taught that "'Partiality should be replaced by universality,'" and "he exhorted everyone to regard the welfare of others as he regarded his own" (p. 410). More recently, around two hundred years ago, Kant ([1785] 1964) similarly proposed the categorical imperative: Act only on that maxim which you can at the same time will to be a universal law. Even more recently, Rawls (1971) grounded his influential theory of justice in Kantian-like impartiality. In what Rawls calls the "original position," he asks us to imagine that we are to be born into a society and do not know what our positions will be: wealthy or poor, healthy or sick, talented or not, and so on. Rawls then asks us to choose basic moral principles under which we would want to live. Rawls contends that once we are placed under this "veil of ignorance"—that is, once we are equally encumbered by not knowing about our own particular qualities—we will chose egalitarian moral principles that are binding on all rational people. Along similar lines, Gewirth (1978) defines morality as "a set of categorically obligatory requirements for action . . . [that apply to everyone] regardless of whether he wants to accept them or their results, and regardless also of the requirements of any other institutions such as law or etiquette" (p. 1).

Over the last several decades, the field of moral development has been particularly shaped by two research programs, one impelled by Kohlberg and the other by Turiel. Both programs have drawn on the above philosophical traditions. According to Kohlberg (1974), for example, a moral judgment involves "a mode of choosing which is universal, a rule of choosing which we want all people to adopt always in all situations" (p. 11). Thus, like Kant, Mo Tzu, and others before him, Kohlberg (1971) viewed the concept of moral obligation in terms of "two formal characteristics of moral judgment, prescriptivity and universalizability" (p. 304). But, departing from the philosophers, we must also keep in mind that Kohlberg understood the concept of moral obligation developmentally.

To explain this point, and the startling implication, consider first an analogy to Piaget's assessment for number conservation. In front of a child of perhaps four to five years of age, lay out in a line six candy M & M's, evenly spaced, say one inch apart. Now lay out an identical but parallel line of six M & M's above the first. Ask the child to count the number of M & M's in each line. The child will count six M & M's in both. Ask the child whether there are more M & M's in the bottom line or in the top line, or the same number in both. The child will say the same number in both. But now, in front of the child, spread the distance between the M & M's in the top line so that overall that line is longer. Now ask the child the same question: whether there are more M & M's in the bottom line or in the top line, or the same number in both. Here is the surprise finding (and a criterion for establishing preoperations): The nonconserving child will say that there are more M & M's in the top line. Of course, one might hypothesize that the child just meant that the top line is longer than the bottom line. So, ask the child some variations of the initial question: "If you were hungry, which line of M & M's would you want to eat, or would it not matter?" Again, the nonconserving child will say the top line. Ask the child to count both lines again. The child does and again counts six M & M's in both. Ask the original questions again; and again the child asserts that there are more in the top line. The point here is that if a child's understanding of number conservation—that the number of objects remain the same no matter how they are arranged spatially—can be undermined by a perceptual "seduction," then the child's cognitive knowledge about number is pseudo knowledge, not necessary knowledge, not logically compelling.

Kohlberg makes a similar claim about moral judgments. Recall from chapter 3 Kohlberg's basic developmental characterization: that moral reasoning proceeds from a focus on punishment (stage 1), to instrumental hedonism (stage 2), to social approval (stage 3), to the maintenance of the social order (stage 4), to social contract, human rights, and universalized principles of justice (stages 5 and 6). Thus only by stages 5 and 6 (postconventional morality) does a real and necessary concept of moral obligation emerge. Kohlberg (1971) writes:

> The individual whose judgments are at stage 6 asks 'Is it morally right?' and means by morally right something different from punishment (stage 1), prudence (stage 2), conformity to authority (stages 3 and 4), etc. Thus, the responses of lower-stage subjects are not moral for the same reasons that responses of higher-stages subjects to aesthetic or other morally neutral matters fail to be moral. (p. 216)

In other words, as long as the necessity of a moral obligation does not stand firm in the face of non-moral conflicts, then what looks like an obligation is not real, but merely a pseudo obligation. Since Kohlbergian research has shown that only a small percentage of adults, and even fewer adolescents, and virtually no children score as postconventional reasoners, the implication is startling: It follows that most adults, even more adolescents, and virtually all children do not make moral judgments. To many people this implication seems counterintuitive, and leads to the question: Can this proposition be true?

Turiel and his colleagues think not. Drawing particularly on Gewirth's (1978) moral philosophy, Turiel and his colleagues have defined obligatory moral judgments as prescriptions that are (a) generalizable, meaning that they apply universally to all people in morally similar situations, (b) not contingent on societal rules, laws, or conventions, and (c) justified by considerations of justice, fairness, rights, or human welfare. In turn, they have used these criteria to differentiate moral issues from those that are personal and conventional. Personal issues refer to those that lie under the jurisdiction of the self. Conventional issues refer to behavioral uniformities designed to promote the smooth functioning of social interactions. This perspective has become known as the "domain perspective" in moral development.

Here is an example of how the distinctions between these social domains—the moral, conventional, and personal—play out in their basic form through social-cognitive interviews with children. Consider a school in the United States that requires children to call teachers by their surname with Mr., Mrs., or Ms. Research typically shows the following pattern of results. The interviewer first asks a prescriptive question, often framed in terms of whether an act is all right or not all right to perform. In this scenario: "Is it all right or not all right to call teachers by their first names?" The student will answer, "Not all right," and often justify his or her evaluation based on an appeal to conventions ("Because that's the way we do things around here"). The interviewer then asks some version of a rule-contingency question: "Let's say that the principal of the school said that it is all right to call teachers by their first names, is it now all right or not all right to call teachers by their first names?" Now the student will answer, "All right." Then the interviewer asks some form of a generalizability question: "Let's say that the principal of a school in another country says that it is all right for students to call teachers by their first names; is it now all right or not all right for those students to call teachers by their first names?" The student will again answer, "All right." Justifications for the two latter evaluations appeal to the relativity of conventions ("If people in authority decide to do things differently, they can").

In the second scenario, instead of asking about a conventional issue, the interviewer asks about a moral issue. Perhaps the interviewer asks about an event that involves unprovoked physical harm, such as when an older child (a bully) starts a fight with a younger child. The student's evaluation for the first question will typically stay the same: "It is not all right for an older child to start a fight with a younger child." But the justifications differ, appealing to justice or human welfare ("It's not fair, because the younger child isn't doing anything wrong, and plus the older child could hurt him"). In addition, the student will say some version of the following: "It's not all right to start a fight even if the teacher and principal make a rule that says it is all right" (rule contingency); and "It's not all right even if another school in another country says it is, and allows it to occur" (generalizability). Justifications appeal to justice and human welfare and are not contingent on personal interests, authority dictates, or conventional practice.

In the third scenario, the interviewer asks about a personal issue, perhaps whether it is all right or not all right for a student to choose his or her own friends. The student will say it is all right, and offer justifications that appeal to the legitimate claims of individual interests when such actions lie outside moral and conventional purview ("It's my choice, and my parents don't have the right to tell me who my friends should be").

More than seventy published empirical studies, in Western and non-Western countries, have demonstrated that children, adolescents, and adults distinguish moral issues from those that are personal and conventional (e.g., Hollos, Leis, and Turiel 1986; Killen 1990; Killen, Leviton, and Cahill 1991; Laupa 1991; Madden 1992; Nucci 1981, 1996; Smetana 1982; Tisak 1986; Tisak and Turiel 1988; Wainryb 1995; for reviews of the literature, see Helwig, Tisak, and Turiel 1990; Smetana 1995, 1997; Tisak 1995; Turiel 1983, 1998; Turiel, Killen, and Helwig 1987). Thus the issue is not whether the data domain theorists have provided is replicable—even critics grant that it is (Campbell and Christopher 1996; Glassman and Zan 1995; Haidt, Koller, and Dias 1993). The issue is how to interpret it. Kohlbergians will continue to say that the obligation is a pseudo obligation because the judgment is not overriding in all situations. I think that is a coherent position, but not the most compelling one. Rather, the domain-specific research has established that children at a relatively early age have concepts that are prescriptive, generalizable, not contingent on societal rules, laws, and conventions, and justified by considerations of justice and welfare. In my view, that is enough to establish a moral obligation.

Discretionary Morality

Recall the incident with my daughter and the homeless man on the streets of Cambridge. Were we morally obligated to give this man some money? Are you morally obligated to give some of your money to Amnesty International? Oxfam America? A local AIDS hospice? Was Mother Teresa of Calcutta morally obligated to spend her life attending to the destitute? Many people would consider these and other situations and say no. But here is a problem. If morality is defined in terms of obligatory requirements, and if such actions are not obligatory, then performing these actions would, by definition, lie outside the moral domain. Yet surely helping strangers in need is moral. What else could it be?

Domain theorists, as I understand their position, might respond in one of two ways. The first response would be to suggest that such actions lie within the personal domain. Granted, such characterizations may be accurate for situations where another person's need is minimal. John and Bill are sitting next to each other. John drops a pencil. Does Bill pick it up for him? Either way, it appears that such "prosocial" behavior says more about personal predilections than morality. But I think domain theorists would agree that this analysis falls short when the actions have substantial benefit to others' welfare. One person says, "I spend my Sundays helping AIDS patients in a local hospice, without pay." Another person says, "I spend my Sundays watching NFL football and reruns of *I Love Lucy*." Are both people engaged in the same category of action? No, even if there are personal motives in the first course of action, if not in all human action, reducing the former statement to the latter runs roughshod over the person's moral concern for the well-being of others.

A second response from the domain approach is more complex. It is to assert that the moral obligation continues to exist, but that it gets coordinated with personal interests in such a way that the moral action is not acted upon. But consider this position in light of my earlier questions. I asked, for example, Are you morally obligated to give some of your money to Amnesty International? Oxfam America? A local AIDS hospice? Was Mother Teresa of Calcutta morally obligated to spend her life attending to the destitute? These are sensible questions. And if concepts of moral obligation are not found in reasoning about such situations, it seems strained to interpret this finding as meaning that an obligation to perform these actions still exists (e.g., "I am morally obligated to spend every Sunday helping out at a local hospice"), but that the obligation is trumped in such situations by personal interests ("but I think I'll watch TV instead"). Rather, it seems more straightforward to say that in such situations the concept of moral obligation is not present. In turn, once that is said, there is the need to broaden the moral domain to include acts that are simultaneously nonobligatory but moral.

In my own research, I have followed this line of reasoning by distinguishing between two forms of moral judgments (Kahn 1991, 1992, 1995). The first is obligatory moral judgments, as defined by domain theorists: prescriptive judgments that are also universalized, not contingent on rules, laws, and conventions, and justified based on moral considerations of justice and human welfare. The second I have termed discretionary moral judgments. This term "discretionary" follows Williams's (1985) view that there may be actions that are "heroic or very fine actions, which go beyond what is obligatory or demanded. Or there may be actions that from an ethical point of view it would be agreeable or worthwhile or a good idea to do, without one's being required to do them" (p. 179). In other words, discretionary moral judgments are those where moral action, while not required of an agent, is nevertheless conceived of as morally worthy based on concerns of human welfare or virtue (see also, Fishkin 1982; Hart et al. 1995; Nisan 1991; Hunt 1987; Urmson 1958).

This distinction between obligatory and discretionary moral judgments emerged in an early collaborative study I conducted on children's conceptions of trust (Kahn and Turiel 1988). We presented children of three age levels with three different situations where trust was violated in a friendship. The first situation violated a moral trust in a friend's honesty: a child lies to a friend in order to get the friend to share his lunch. The second situation violated a trust that a friend would provide emotional support in a time of need: a child does not forgo a personal interest (watching television) to comfort a friend through an emotional difficulty. The third situation violated a trust that a friend would adhere to social conventions in an appropriate context: a child fails to dress appropriately to his friend's birthday party at a fancy restaurant.

Results showed many ways in which these different types of trust violations differentially affected children's friendships. For example, it was found that the majority of children would feel their friendship was diminished in intimacy by the violations in the first two situations (honesty and emotional support), but not the third (conventions). Age-related differences across situations were also found. For example, at ages 6 and 7, children conceived of friendship in particularistic terms, wherein friendship relations depended on particular events and personal predilections ("because she lied to me"; "because I wanted him to play with me"). At ages 8 and 9, children held to a more generalized view of friendship based mainly on the frequency and magnitude of positive and negative interactions (e.g., "She would still be my friend as long as she didn't do it a lot of times"). At ages 10 and 11, children integrated positive and negative interactions within the context of the reciprocity of the relationship ("All those years of being friends and stuff, why let it go away in one day"). I shall come back to this developmental finding in chapter 8, where I present an intriguingly similar shift in children's conceptions of what it means to live in harmony with nature.

But for purposes here I want to highlight one finding in particular: a greater number of children negatively evaluated the violation entailing deception (97%) than the violation entailing emotional support (61%). The justifications in the first instance focused on appeals to fairness, obligation, and maintaining or establishing relationships. The justifications in the second situation appealed usually to the emotional well-being of others and the personal choice of the actor. In other words, although our study was designed for other purposes, what I think was going on was that children viewed deception as a violation of a moral obligation. In contrast, many children did not believe that providing emotional support was a moral obligation, but still a desirable course of action based on concern for others. This latter reasoning begins to look a lot like what I have termed discretionary morality.

These results led me to pursue the distinction between obligatory and discretionary moral judgments more directly. In particular, I did so while also exploring a potential asymmetry between positive and negative morality. Positive moral situations refer to the "do's" of the moral life (e.g., practice charity), and negative moral situations refer to the "do not's" of the moral life (e.g., do not steal).

Here is what I mean about this potential asymmetry. Let us go back to the homeless man whom my daughter and I encountered on the streets of Cambridge. Let us also say for the moment that I believed I had a moral obligation to aid this man in some way. But as I asked earlier, what is my response when I meet a second homeless person who asks for money? Or a tenth person, or a hundredth person, or a ten thousandth person, as in, say, an African famine relief appeal? At some point I will likely say, "Enough is enough, I have my own needs to take care of and my own personal interests to pursue." That is, there are personal costs involved in helping others, and these costs presumably affect conceptions of whether one is morally obligated to help. In contrast, when I say something like, "I have a moral obligation not to steal from others," I mean that to apply not only to one or two or a hundred other people whom I might meet, but to all people. This is so even though with each person from whom I do not steal my "costs" increase (loss of potential revenue). Thus it would appear that concepts of moral obligation are more stringently applied to negative than to positive morality, where by stringency is meant a property of a moral view being very demanding within its area of application (Scheffler 1986, 1992).

Toward investigating these issues, I interviewed seventy-two children in grades 2, 5, and 8 about three brief stories that controlled for the degree of personal cost—low or high—incurred for performing a moral act (Kahn 1992). The first story entailed positive morality: a protagonist faces the potentially moral action of giving a small portion (in the low-cost condition) or all (in the high-cost condition) of his lunch money to an economically impoverished woman and her four children who have not eaten in a day or two. The second story also entailed positive morality, but here the protagonist faces the potentially moral action of giving a small portion (low cost) or all (high cost) of the hundred dollars he won in a raffle contest to (again) an economically impoverished woman and her four children who have not eaten in a day or two. Finally, the third story entailed negative morality: a middle-class mother (low cost) or economically impoverished mother (high cost) considers taking (stealing) the lunch money of a child. Thus in each story the low-cost condition was weighted toward an obligatory orientation and the high-cost condition toward a discretionary orientation.

The results provided evidence that children as young as second grade make distinctions between moral acts that are obligatory for a moral agent to perform, and moral acts that are left to the moral agent's discretion. This distinction, in turn, supported the proposed asymmetry between positive and negative morality. For example, in the low-cost conditions about 40% of the children judged the protagonist in the first two stories morally obligated to give money to the needy family, while 100% of the children judged the protagonist in the third story morally obligated not to steal. In contrast, in the high-cost conditions virtually none of the children (around 2%) judged the protagonist in the first two stories morally obligated to give money, while the large majority (87%) still judged the protagonist in the third story morally obligated not to steal. These results support the proposition that while obligations can cut across children's reasoning about positive and negative moral acts, it is also the case in many situational contexts, particularly those that pivot on agent's cost, that children conceive of negative morality more stringently than they do positive morality.

Now, it may be asked: In what way were these morally discretionary judgments to be distinguished from strictly personal judgments, such as a judgment for which flavor ice cream to choose for dessert? Toward making this distinction, two criteria were used in consort. The first criterion drew on Williams's (1985) proposal that one obvious way to assess the moral status of a discretionary act is to assess whether the act would be greatly admired or well thought of. This criterion harkens back to the Aristotelian concept of the "good" noted earlier, and was assessed in terms of praiseworthiness: whether children thought the protagonist of the story should be praised for performing the positive act. Results showed across all four positive conditions that of the children who viewed the positive acts as discretionary, the large majority (over 90%) provided praiseworthy evaluations. The second criterion drew on children's justifications for praising. Two results are of interest. First, across all four positive conditions the large majority (over 86%, with an average of 94%) of justifications children provided for praising entailed concern with others' welfare or virtuous character. Second, of the children who said the low cost positive acts should be performed (but viewed the acts to be discretionary), the large majority (over 90%) of their justifications entailed concern with others' welfare. Given that concerns with human welfare and virtue are central to moral discourse and moral theory, and taking such justifications in conjunction with the praiseworthy evaluations, the results help establish children's discretionary judgments as moral.

Based on the criterion of praiseworthiness, results also showed that children accorded the most praise to performing positive discretionary acts, somewhat less praise to fulfilling positive obligations, and even less praise or no praise to fulfilling negative obligations. Such results add specificity to the intuitive notion that praise applies more readily to positive than negative morality. However, the results need to be understood within the context of what children meant when they accorded praise. In the interview children were posed with questions pertaining to whether they would praise the person for performing the act. Results showed that younger children provided more act-centered justifications, while older children more agent-centered justifications. In other words, while younger children readily engaged in dialogue about this issue, they consistently talked about praising the act and not the actor. I will come back to this issue in chapter 8.

One final result from this study bears on issues that will arise in subsequent chapters. There has been some heated controversy in the field on whether there exist gender differences in moral development. This controversy arose from Gilligan's (1982) initial charge that while men and boys are primarily oriented toward justice, women and girls are primarily oriented toward an ethic of care. Three results did not support this proposition. First, no gender differences were found in the use of welfare (carelike) and justice justifications. Second, as a group and individually, children often provided both justice and welfare justifications in supporting a particular evaluation. Third, given the proposition that females more than males are oriented to an ethic of care, it could be reasonably expected that girls more often than boys would have said the protagonist should give money to the starving family, particularly in the high cost conditions where the stimuli were less weighted toward moral obligation. For, particularly in these latter conditions, feelings of care and not considerations based on duty or justice would presumably motivate helping evaluations. However, the evaluation results showed no effects of gender. These results are in agreement with a large body of research in the field which has by and large not supported Gilligan's initial claim (e.g., Garrod, Beal, and Shin 1990; Lourenço 1991; McGillicuddy-Deisi, Watkins, and Vinchur 1994; Smetana 1984; Smetana, Killen, and Turiel 1991; for reviews of the literature see Killen 1996; Turiel 1998; Walker 1984, 1986, 1991).

Part of the difficulty in examining where and how gender differences do occur in moral development may be because of the structural interrelatedness of what in the literature has to date been treated largely as diverse and distinct moral constructs. Two results in particular address this issue. In this study, younger children emphasized welfare concerns and older children justice. These results are consistent with previous studies by Davidson, Turiel, and Black (1983) and Walker (1989), and support the proposition that justice concepts build on concepts of welfare (see also Arsenio 1988; Arsenio and Lover 1995). More direct support for this proposition comes by reconceiving one of the justification coding categories, welfare-in-compensation. This category—which, like justice, was used more by older than younger children—had been analyzed and reported as a welfare category, as the welfare needs of the protagonist and family were set in some compensatory balance (e.g., the protagonist should give the money "because he would still have something for breakfast and dinner and the lady would still have all of the money to buy food"). However, this category can also be viewed as welfare considerations organized by an equilibratory structure, since the very balancing of two distinct welfare claims may reflect a form of justice, though framed so as to highlight human needs. This is not to say that justice replaces welfare, but necessarily incorporates it. Thus this analysis moves toward understanding how both constructs depend and draw upon each other ontogenetically (Boyd 1989; Killen 1996). In subsequent chapters, I build on this type of analysis not only to speak to issues of gender differences (or lack of) but also in understanding the development of biocentric reasoning.

Conclusion

It is both obvious and true that people can differ in their moral convictions. Indeed, following from my line of reasoning in the previous chapter, it follows that a single individual can hold multiple if not contradictory moral judgments. Thus, as I see it, the research challenge is how to investigate morality broadly enough to be sensitive to the richness in the moral life by taking seriously diversity of moral judgments, yet precisely enough to delineate the relations between the constructs when they occur. In this chapter, I have offered one such approach based on a framework that distinguishes between obligatory and discretionary morality. We are now poised to apply this framework to environmental content.

5

Structural-Developmental Methods

This chapter provides the reader with the necessary methodological background to understand more fully the empirical research in subsequent chapters. In addition, I believe this methodology has far reaching applications for others interested in pursuing their own related lines of inquiry. Thus this chapter also aims to provide researchers from other fields with enough detail so as to draw on these methods themselves.

In brief, the methods involve what is called the semistructured interview, which was pioneered by Piaget (e.g., [1929] 1960, [1932] 1969, [1952] 1965) and has been elaborated upon by a large number of more recent researchers (e.g., Colby and Damon 1992; Damon 1977; Eisenberg 1989; Ginsburg 1997; Helwig 1995; Keller and Edelstein 1991; Killen 1990; Kohlberg 1984; Laupa 1991; Lourenço 1990; Miller 1994; Berkowitz, Guerra, and Nucci 1991; Oser and Althof 1993; Rest and Narvaez 1991; Rogoff 1990; Saxe 1990; Selman 1980; Smetana 1995; Thorkildsen 1989; Tisak 1986; Turiel 1983). On a basic level, the interview methodology simply involves asking questions to understand a person's reasoning and values—which, of course, is nothing new. Socrates was fond of asking such questions himself. But as a formal research methodology, it presents challenges. I will discuss seven of these challenges in the order they would likely be encountered in conducting a study: constructing the interview, enlisting subjects, interviewing, generating a coding manual, coding the data, reliability coding, and data analysis.

Constructing the Interview

Let us say you wanted to conduct a study on children's reasoning and values about "X" and you decided to employ the interview methodology.

What should you ask children? One option is to ask children directly about the topic. "What is X?" "How do you reason about X?" "Can you give me an example from your own life of when you encountered a problem that involved X?" There is some merit to this direct approach. Certainly it gives children the opportunity to define the problem in their own terms. But quickly you would discover that it comes up short. Perhaps the greatest problem here is that children have concepts about many things on which they cannot directly reflect. For example, in the trust study I described in the last chapter (Kahn and Turiel 1988), we had initially piloted the question, "What does it mean to trust a friend?" We discovered that children, especially in the first grade, had very little to say in response to this question.

Rather, if you choose to use the interview methodology you will usually be better served by employing a less direct, but more substantive, approach. As is common in social cognitive research, you could interview children about a hypothetical situation, or a common every day event in their lives, or a task you have asked them to solve, or a behavior in which they had just that hour engaged. But no matter what your choices here, the important point is a priori to conceptualize what the topic entails, if possible demarcate its boundaries through formal criteria, and at a minimum employ issues or tasks that engage children's reasoning about the topic under investigation. For example, in the above trust study, we drew in part on the philosophical literature to propose that trust be differentiated from mere reliance on things, and be conceived of as expectations within social if not moral contexts. We then created scenarios that violated such expectations and about which even young children had plenty to say. Kohlberg followed a similar approach. Drawing substantively on moral philosophy, Kohlberg proposed that morality centrally concerns issues of justice and procedures for how to adjudicate competing claims. Instead of asking children, "What does it mean to be just?", Kohlberg constructed dilemmas that pitted competing claims against one another, and he used these dilemmas as the basis for an interview. The reason philosophy (and not only other relevant empirical literature) can so often help with constructing the interview is that philosophers make it their business to think clearly about many of the issues that social scientists want to investigate empirically. Indeed the relation between philosophy and psychology too often goes overlooked in both fields (see Campbell and Bickhard 1992; Campbell and Christopher 1996; Piaget 1970, 1971a, 1971b, 1971c).

Below I list seven categories of questions that can help structure an interview. The specific examples come from the Houston child study (chap. 6), wherein we interviewed children across grades 1, 3, and 5 from an inner-city black community on their environmental views and values. (For more detail, see Appendix A for a complete set of our structured interview questions for the Portugal study.)

1. Questions about the general topic (e.g., "Do you think about nature and the environment much? If so: What do you think about? What kinds of issues in the environment concern you?"). Notice that this question allows the child to focus on relevant issues of his or her choice. It can provide some insights, and has the additional advantage of leading the child gently into what will become a more focused interview.

2. Questions about specific content-based aspects of the topic (e.g., "Are parks or open spaces an important part of your life? If so, how?"). Notice that a priori we had conceptualized parks and open spaces as an important and conceptually relevant part of nature with which the child likely has had direct experience.

3. Questions in relation to a hypothetical scenario (e.g., "Let's say someone throws his trash into the bayou. Is that all right or not all right? Why?"). Notice that this hypothetical situation involves a plausible if not common real life situation for the child, and thus particularly helps the first-graders to feel engaged. Equally important, it sets up other questions that allow for the careful manipulation and exploration of theoretically-motivated constructs such as rule contingency and generalizability.

4. Questions about the child's behavior (e.g., "Does your family talk about the environment much? If so: What kinds of things do you talk about? How does a typical conversation about the environment get started?"). Notice that through an interview it is possible to get a tentative measure—based on self report—of behavior, to have as a piece in a broader analysis.

5. Questions about the child's knowledge ("Do you know about the problem of air pollution? What is it?")

6. Questions about the child's beliefs ("Does the problem of air pollution affect you in Houston? If so, how?").

7. Questions about the child's values (e.g., "Does it matter to you if birds got harmed by throwing trash into the bayou? Why?").

Not all seven categories of questions need to be employed in every study. But this list can act as a quick check, or helpful prod, while constructing an interview. It is also worth pointing out that in hindsight I still wonder about the way I specifically phrased some of the above questions. Our usual way of phrasing a question takes the form of number three above—"Is it all right or not all right . . . ?"—where the interviewer offers the participant both the affirmative and negative choice. But in the phrasing of question seven, for example, the interviewer offers only the affirmative choice: "Does it matter to you if birds got harmed . . .?" At the time, this phrasing made sense to us because we had been told by some prominent members of this black community that the children of their community had virtually no interest in environmental issues, and thus we wanted to frame questions that could most easily tap children's possible engagement with nature. The questions worked as we had hoped, but they also opened us up to the charge that the interviewer led a participant to a judgment. It was a difficult call, one of many that researchers make in constructing an interview.

A decision also has to be made about how long to make the interview. Often the decision is constrained not by one's research questions, which may be virtually unlimited, but by the resources available for the research (e.g., time and funding opportunities). An additional concern is the attention span of the participants. Interviews with college students can last as long as two hours. Although the interviews also can be broken into two parts conducted on different days, my own inclinations are, whenever possible, to conduct an interview in one session so as to maximize each interview's coherence and continuity. With younger children, shorter interviews are needed. It is also the case that younger children often have less to say during an interview, so that sometimes, with the same interview protocol in a developmental study, age and interview length increase together. But let me be clear that in my experience even six- and seven-year-old children can stay engaged during an interview for up to forty minutes. Success depends largely on two things. First, one needs to construct the interview in ways that engage children in a lively discussion, rather than disengage them on an abstract set of questions to answer in a testlike fashion. Second, one needs to interview well, which I will say more about shortly.

After constructing an interview, it is then essential to try it out—to pilot it—with some participants. At that point it will become clear that some things are working and other things are not, and modifications in the interview are made accordingly. Typically, colleagues and I go through several rounds of piloting before we settle on a final interview protocol.

Enlisting Participants

In most of our studies, we have sought to interview a sample of a population which would by and large be representative of some larger population. We are all too aware, however, of our methodological short-comings. Our research with children, for example, requires the written consent of each child's parent. How do we get such consent? Typically, cooperating teachers send parent permission forms home with their students. These forms, in turn, have been scrutinized and amended by university committees that seek to protect "human subjects"—meaning, that the forms are laden with information that is useful to the literate and interested parent, and usually ignored by other parents. Very roughly, we tend to garner a 50% return rate on parent permission forms, including a very small percentage of parents who explicitly do not give consent. Thus, right here, a substantial preselection occurs on dimensions which are difficult to assess: perhaps involving literacy, a coherent home environment, parental interest in their child's education, parental interest in the study, the child's interest, and so forth. Such problems reflect the difficulties of most social science research, and lead to some caution in generalizing from one's data.

Interviewing

In all of our studies, we interviewed our participants (including parents) on school grounds. With children, we arrange times that work for the cooperating teachers and a child, meet the child in his or her classroom, and then move to and conduct the interview in an empty, quiet room. The interviews are tape-recorded. If this recording technology appears unfamiliar or intimidating to the child, it can help to have the child work the machine, try a sample twenty seconds of recording ("hello, testing, today is March 26th, and it's a beautiful day, and now you say something . . . "), and then play it back. It also helps to chat with the child informally, to help establish rapport. I remember one seven-year-old boy who, before the formal interview, talked about his special lunch, and his dialogue went something like this: "It's very, very special, and my mom makes it for me because she loves me. It's so special, but I don't know what it's called. It has bread. And on one side there is peanut butter. And on the other side there is jelly." Interviewing has many rewards.

If setting up the interview situation is relatively straightforward, interviewing itself is not. Piaget ([1929] 1960) says this:

> [It] is so hard not to talk too much when questioning a child, especially for a pedagogue! It is so hard not to be suggestive! And above all, it is so hard to find the middle course between systematisation due to preconceived ideas and incoherence due to the absence of any directing hypothesis! The good experimenter must, in fact, unite two often incompatible qualities; he must know how to observe, that is to say, to let the child talk freely, without ever checking or side-tracking his utterance, and at the same time he must constantly be alert for something definitive, at every moment he must have some working hypothesis, some theory, true or false, which he is seeking to check. . . . When students begin they either suggest to the child all they hope to find, or they suggest nothing at all, because they are not on the look-out for anything, in which case, to be sure, they will never find anything. (p. 9)

Thus, in interviewing, Piaget advocates for the middle course. On the one hand, the researcher does not allow the child to ramble aimlessly. On the other hand, the researcher does not want to prod the child into saying things that the child thinks the researcher wants to hear. Young children in particular are susceptible to such implicit coercion, and measures must be taken to guard against it. One important measure occurs while constructing the interview, where one wants to build into the protocol a neutral wording of the questions. An engaged neutrality must also be part of the interviewing process.

Moreover, if at any point in an interview that you feel that you are not getting the child's own convictions, you can offer the child a counterexample. You might ask: "You know, another child told me that they thought it would be all right to throw trash in the bayou, because it's just one little piece, and it's not going to do any harm. What do you think about what that child said?" Such counterexamples legitimate a differing conviction in two ways: by offering a compelling reason, and by putting the evaluation and reason in the words of another child. Indeed you often arrive at compelling counterexamples by drawing on what other children have said during pilot interviews. For the child who does not change convictions (and in my experience children usually do not), counterexamples often serve to prod the child to develop his or her reasons more fully. In turn, for the child who changes convictions in response to a counterexample, it is then possible to offer a countercounter example—which is often simply the child's original conviction—and see if you get a shift back. But such techniques always must be used judiciously so as not to intimidate, bewilder, or otherwise alienate the child. It is also the case that, when theoretically motivated, counterexamples can be incorporated systematically as standard interview questions (see, e.g., questions 35 and 36 in Appendix A).

In my view, one of the key components of successful interviewing is to seek sincerely to understand the participant's understandings about the issues under investigation. "I don't quite follow, help me understand that better." "Well, how do you understand this problem?" Children usually enjoy being interviewed. Too often, schools ignore their viewpoints. Interviews offer children a chance to be understood about issues they care about. Adults, too. In the interviews we conducted with the Black parents in Houston (chap. 7), for example, we not uncommonly received supportive comments at the end of the interview. Here are three (with the respondents' words in italics):

Anything else you want to say or ask me? *Well, I enjoy, enjoy this.*
Well that's our last question. *Really?* Yeah. *I was ready, now see I was ready to go on there.*
Well, Brenda, that was my last question. *OK. What you talking now, that was a good conversation.*

It is such participant engagement that increases a researcher's trust in the resulting data.

One issue that often arises when people learn to interview is knowing how, and how far, to pursue the participant's justifications. This issue can arise particularly when interviewing young children and children with little verbal agility. What happens after you ask, "Why?", and they say, "I don't know." One approach is just to paraphrase what the child has said so that he or she can hear it, and that sometimes helps the child's reflection: "You said X and I'm just wondering why you think that?" But it is less common that children provide no justification and more common that their justifications are undeveloped. When this occurs you can draw on a range of possible questions: "Can you say more about that?" "How do you mean that?" "Earlier you said X, is that what you mean here, or is it different? How so?" "Can you give me your own example of that?" "Has that ever happened to you? Tell me about it. Well, having said what you said, what do you think about the other situation we've been talking about?"

Here is where experience counts. One way of knowing that a justification is undeveloped is if you have previously worked on coding similar data and run into problems in interpreting ambiguous responses. As a coder you often say, "Drat, I wish the interviewer had asked X, and then I would know what this child means." Many "drats" later you are in a strong position to interview because you know better what you need to know to understand the participant's position. This is part of what Piaget means in the above passage when he says that the interviewer "must constantly be alert for something definitive, at every moment he must have some working hypothesis, some theory, true or false, which he is seeking to check." Accordingly, it also helps to sit down with the questions for which you plan to pursue systematic justifications, and do the best you can in conceptualizing the range of possible responses, what undeveloped responses would look like, and what follow up questions could clarify the ambiguity.

In our studies, ideally we stopped pursuing justification data only when (a) one or more coherent justifications were provided, (b) possible conflicts between justifications were probed out ("Earlier you said X and now you say Y, can you help me understand this difference?"), and (c) the participant said she or he didn't have any further reasons. But obviously this ideal was not always met. To keep the participant engaged with the interview process (and thus strengthen internal validity and how much we trust the data we do get) the interviewer sometimes had to give ground and not pursue justification data to its end. Perhaps in the beginning of the interview a participant feels apprehensive, and the interviewer decides not to push too much in order first to establish rapport. Perhaps a participant had already developed ideas extensively on a question, and felt too much repetition in the next question. Perhaps toward the end of the interview, the interviewer can feel the participant's concentration waning, and moves ahead more quickly. All of these judgment calls, and more, an interviewer makes on the spot.

Generating a Coding Manual

After interviewing, all of the audiotapes are transcribed. A typical transcript will range from 15 to 30 single-spaced typed pages. Take an average length times the number of participants, and you end up with a stack of perhaps 2000 pages of qualitative data sitting on your desk. What next? It is time to generate a coding manual. By a coding manual I mean a systematic document that explicates how to interpret and characterize (and thereby "code") the qualitative data. Coding manuals vary enormously in their underlying approaches and physical length, which can range from several pages for simple studies to book-length documents for studies that follow an extensive research program (like Kohlberg's). Given such variations, I shall limit this discussion to some key ideas and how they have been implemented in coding manuals colleagues and I have generated. I have also included in Appendix B the justification coding section of the coding manual Lourenço and I generated for the study we conducted in Portugal, where we interviewed students in grades 5, 8, 11, and college.

Generating a coding manual takes time. Often you begin by sitting around a table with one or two colleagues, reading the first interview aloud. At each segment you stop, and someone says, "This is what I think is going on, what do you think?" Sometimes the interpretations are simple, virtually a tautology. For example, if a participant answers an evaluative question by saying, "I don't believe that's all right," then you can just go right ahead and characterize the evaluation as "not all right." But at times even evaluation data can require your thought. Let us say that a participant says, "Ummm, I'm not sure, but, yeah, I tend to say that's not all right." Now you have a decision. You could broaden your manual to include three evaluation categories: all right, unsure, and not all right. Where then would you fit the above evaluation? In my interpretation, the participant leans more toward the "not all right" code than the "unsure" code. But to get even a better characterization, you could broaden your manual further to include five evaluation categories: all right, unsure but substantively leaning toward all right, unsure, unsure but substantively leaning toward not all right, and not all right. At this point the above evaluation would get squarely placed in the fourth category: "unsure but substantively leaning toward not all right."

You can see how it goes: as you work with more of your qualitative data, you can arrive at greater specificity of characterizations. But you have to be aware that you are also after ways of understanding and later speaking about the data in broad terms, and that unnecessary details will only get in the way. Thus, in the above situation, it may not be important to have so many separate evaluation categories, especially if most of them are seldom used. If this is the case, as it usually is, then you can simply settle upon "all right" and "not all right," and then fall back on the code "uncodable" when you encounter the occasional response that strains the boundaries of either category.

When you turn to the justification data, the complexities increase dramatically. For instead of trying to characterize accurately one or two sentences revolving around an evaluation, you might have a segment of two or ten or twenty sentences wherein a participant provides multiple justifications for his or her evaluation. Still, the challenge remains the same: to generate a system of coding the data that at once provides as much conceptually relevant detail as possible, while providing an elegant overarching framework by which to understand and later speak about the data as a whole.

One of the key ways of meeting this challenge is to build on the idea of hierarchical classification: that some ideas are subsets of other ideas. Let us say you read the following justification for why it is not all right to pollute the water: "because if the water is dirty, I might get sick." You first interpret this justification in terms of a focus on human welfare (the idea of human sickness). Then you decide "welfare" would make a plausible coding category because welfare, after all, is central to moral theory (chap. 4). Then you read some more justifications and realize that some participants focus not just on their own welfare (as above) but on the welfare of others ("air pollution goes by and people get sick; it really bothers me because that could be another person's life"). You then think that this distinction is worth developing because it is fundamental to sociobiological theory (chap. 2), developmental theory (chap. 3), and moral theory (chap. 4). Thus, under the welfare category, you create two subcategories: welfare of self and welfare of others. Then you discover that under each of these subcategories, participants provide specificity about the type of welfare. A participant might focus on physical welfare ("because I might get a disease"), material welfare ("because a car could get a flat tire"), or psychological welfare ("because I'd be sad"). These three categories then become sub-subcategories. Then you come upon something big. You realize that while some participants focus on the welfare of humans, other participants focus on issues of fairness for nature ("Because it's not fair to the animals who died"), or respect for nature ("[You need] to respect lower life forms and respect animals when you see animals"). That leads you to recognizing that perhaps there are two broader forms of reasoning, one human centered (anthropocentric) and another nature centered (biocentric). Those two categories then become overarching, with welfare and fairness embedded within. And so it keeps going and going, as you fill out and modify a coding manual while reading through the interviews, until you have mined the data completely.

That is what generating a coding manual might begin to look like if you were doing it entirely from scratch. But by virtue of having constructed a philosophically guided interview, you probably already have conceptualized some categories. These categories, in turn, do not so much emerge from the data, as above; rather, with the data in hand, you seek to confirm and then refine them. Moreover, most coding manuals draw a priori from earlier ones, in terms of both establishing individual categories and an overarching framework. For example, the coding manual presented in Appendix B drew on Kahn (1997b) and Kahn and Friedman (1995), which had drawn on Kahn (1992), which had drawn on Kahn and Turiel (1988), which had drawn on Davidson, Turiel, and Black (1983). This sort of "lineage" is common in social cognitive research (compare Arsenio 1988; Helwig 1995; Killen 1990; Laupa and Turiel 1986; Nucci 1985; Turiel, Hildebrandt, and Wainryb 1991; Wainryb 1991). How much one uses of a previous coding manual depends on the earlier manual's comprehensiveness and quality, and how similar your current research topic and theoretical commitments are to those embodied in the earlier manual.

As the coding manual takes shape, you will likely discover that some of your qualitative data can be coded in several different ways. What should you do? The answer depends on how the problem arose. Here are three lines of inquiry to consider.

First, the problem can arise simply because the segment in question contains two or more independent justifications. A participant might say, for example, "It's not all right because people might die and it's not fair to the animals." The initially key word here is the conjunction "and," which suggests two separate justifications: in this case, using the above coding categories, an anthropocentric welfare justification and a biocentric justice justification. One approach some researchers have taken in these sorts of situations is to code only the dominant justification. This approach has the advantage of setting up for later, if not rendering possible, many inferential statistical analyses. But my own approach is to give some ground on the statistics and code every justification that can be legitimately discerned. One reason is because often (as above) it can be arbitrary what one calls a dominant justification. More important, by coding multiple justifications I think you attain a fuller and more balanced characterization of a person's reasoning.

Second, the problem can arise because two categories are conceptually intertwined. This problem occurred in the previous chapter, wherein I described my study on children's obligatory and discretionary moral judgments. In that study, I labeled one of the justification categories "welfare in compensation." Recall that I had analyzed and reported this category as a welfare category, as the welfare needs of a protagonist and a family were set in a compensatory balance (e.g., the protagonist should give the money "because he would still have something for breakfast and dinner and the lady would still have all of the money to buy food"). However, I noted that this category could also be viewed as welfare considerations organized by an equilibratory structure, since the very balancing of two distinct welfare claims may reflect a form of justice, though framed so as to highlight human needs. In such a situation, it does not seem quite right to code the reasoning as two independent justifications (one for welfare and another for justice). Thus I picked the category that seemed dominant (welfare) and went with that, while retaining the idea of the interconnections between welfare and justice, which I could then (and did) address qualitatively in a discussion.

Third, the problem can arise when there is more than one legitimate way to code your data. In this situation, it is not a problem so much as a choice. Remember that the coding categories are driven not only by your data but by your theoretical commitments and research questions. Let us say that in a study you foremost want to understand better the relations between welfare and justice reasoning on environmental issues. In this case, you might well decide to make welfare and justice your most overarching categories, and then subsume anthropocentric and biocentric considerations within each of them. Here I cannot overstate the importance of choosing your overarching categories carefully. For while you will attain detailed levels of specificity in a coding manual, which is good, you will often want to collapse some of the categories at a later time and speak about what the data means on a more general level. That all-important move will be both shaped and constrained by your choices here.

One last point. Ideally you would like to generate a coding manual from one set of data and then apply the manual to an entirely new set of data, so as to verify a replicable effect. But that approach demands more time and resources than most researchers can afford. Still, something can be done in this direction. Namely, before you ever look at the interviews, randomly split your data set in half. Set one half aside, and then develop your coding manual with the other half. Then when it comes time to code the data, apply the manual to the entire data set. By this means you verify what can loosely be referred to as internal replicability (compare Thompson 1996).

Coding the Data

With the interviews completed and a coding manual in hand, the next step is to code the entire data set. In our studies, the coder has usually been someone involved in generating the coding manual. Some of my colleagues can code well a 20-page interview in two hours or less. It usually takes me longer. The important point is not how long it takes but that the coder brings careful thought to each and every code.

Reliability Coding

After coding the interviews, it is important to establish the reliability of your coding system by having a second coder recode a portion of your data. In each of our studies, a second coder recoded about one quarter of the data, spanning three types of responses: evaluation responses (e.g., all right/not all right; aware/not aware of environmental problems; matters/does not matter that insects would be harmed), content responses (e.g., animals, plants, garbage, water pollution, and air pollution), and justifications for the evaluative responses (e.g., an appeal that animals have rights). Intercoder agreement typically fell around the following percentages: evaluations, 95%; content responses 80%–85%; and justifications, 75%–80%. These reliability percentages are representative for published peer-reviewed research of this type. The reason for the comparatively lower percentage of intercoder agreement for justifications is not hard to understand. As described earlier, there is tremendous complexity in justification coding. Indeed, to achieve the numbers that we do, we—and others in this line of research—usually collapse the justification categories to the dozen or so broadest categories that we subsequently summarize in our publications. Finally, Cohen's Kappa has been used as a statistical test of interreliability between coders. In each of our studies, statistical significance was attained at the .05 level.

Data Analysis

Semistructured interview data lends itself well to both qualitative and quantitative analysis. At this point in the study, little needs to be said about qualitative analyses, since most of your analyses will already be done. They lay in the coding manual. In addition, there is now an opportunity to go back to the data and ask: What has been lost due to the formal systematization imposed by a coding system? There are always things of interest. I noted one above for an earlier study of mine, where—in the "welfare-in-compensation" justification category—I lost some of the connections between welfare and justice reasoning by subsuming the category under welfare. In publishing this research, I was able to provide a qualitative discussion of this issue. Remember, too, that in presenting qualitative interview data, you will rightfully encounter a critic who will say something like, "How can I trust that what you as the researcher see in the interviews are the same things I and other people would see?" To a large extent, this concern is addressed by presenting a summary of the coding manual (coupled with illustrative responses) and a measure of intercoder reliability. It also helps to flesh out more fully your interpretations of key justifications. As much as possible, the idea is to put the reader in a privileged position to question or reinterpret your interpretations. Throughout this book, I approach it this way: "Here is an example of my qualitative data; here is how I interpret it; what do you think?"

Even if the reader agrees with the interpretation, another question arises: How many children in the population provided such reasoning? Descriptive statistics will answer such questions. Tally all the codes. Then look for trends in the data, and engage in a limited amount of post hoc analysis by examining what effects collapsing various categories has on the numbers. Then, at the point you ask questions about the likelihood of having gotten certain effects by chance, you make the move to inferential statistics.

Here there are at least three approaches in the field of how to proceed. One approach is to employ nonparametric statistics, including Fisher's exact, chi-square, McNemar, Cochran's Q, Friedman, and Kendall's tau. The advantage here is that the distribution assumptions of these statistical tests are met. But the disadvantage is that you end up with an enormous number of separate tests, which—besides lacking elegance—makes it difficult to assess interaction effects and know how to control the alpha level. A second approach is to convert the nonparametric data to a normal distribution, and then handle the data by more conventional means, such as analysis of variance. The problem here is that this conversion may fundamentally violate statistical assumptions (Marascuilo and McSweeney 1977). A third approach is to engage in log linear modeling. This approach has worked well for some researchers (Tisak

1986, 1993; Tisak and Ford 1986; Tisak, Tisak, and Rogers 1994), but such models require larger sample sizes than is often practical to obtain when interviewing.

In our research, my colleagues and I largely employed nonparametric tests. Sometimes we controlled the alpha level (at the .05 level) by splitting it across some conceptually relevant subset of tests. But since most of our studies sought to extend rather than confirm previous research, adequate attention also had to be paid to the problem of Type II errors (false negatives) along with Type I errors (false positives) (Huberty 1987; compare Marascuilo, Omelich, and Gokhale 1988). Thus we often went ahead and conducted each test at the .05 level, but then treated cautiously any single finding of statistical significance. When appropriate, we converted categorical data to score data and then analyzed it by t tests. Moreover, when we believed we had sufficient grounds, we submitted justification data to arcsine transformations and then performed multivariate analysis of variance and analysis of variance.

In subsequent chapters, I largely avoid statistical terminology to enhance readability. The statistical details for many of our studies have been reported elsewhere, and I encourage readers with more specialized statistical interests to turn to these other publications.

Conclusion

In sum, our methods for the research in the subsequent chapters proceeded as follows. We generated theory guided interview questions. We asked these questions of each participant, while allowing ourselves the freedom to follow the questions in different directions so as to tap each participant's individual understandings. Interviews were conducted one-on-one, tape recorded, and then transcribed. Analysis of the transcripts proceeded by many readings of half of the data set, as we sought to understand the diversity and meaning of responses. Then for each study there was a long process of generating a systematic and philosophically and empirically grounded means for coding the qualitative data (a coding manual). The entire data set was coded, and intercoder reliability was assessed. Finally, the data were analyzed, qualitatively and quantitatively.

By endorsing these methods, I do not mean to imply that there are not other valuable methodological approaches. There are. And, depending on one's question, some serve better or worse. As a point of contrast, Kellert's research on people's attitudes and values of nature (described in chap. 1) is based on surveys. Surveys more easily allow for large sample sizes, and thus provide greater confidence in the generalizability of one's findings. Yet Kellert (1996) also readily notes that surveys "represent a blunt instrument for exploring the complexities of how people perceive nature" (p. 38). What prospers with structural-developmental methods is the depth of understanding people's reasoning and the developing organization of that reasoning.

As another viable alternative, Kempton, Boster, and Hartley (1996) have employed a two-pronged methodology. In a compelling research project, they investigated the environmental values of diverse constituencies in the United States by first conducting 46 semistructured interviews. Then they abstracted pivotal and prototypic responses from their interviews to include in a fixed-form national survey. Thus they aimed to get the best of both traditions: the richer, textured insights that arise through interviews and the better generalizability that arises through surveys. I think they succeeded—but I would note two points. First, the survey methodology still limits the claims about which they can generalize. It is one thing for a participant to be able to generate an idea (as in an interview), it is another thing merely to agree or disagree with an idea generated by somebody else (as in a survey). Second, Kempton, Bolster, and Hartley did not seek to codify their interview data, in the sense of systematically coding it and thereby making it amenable to quantitative analysis. Their decision met their needs, but they had other options, too.

The larger idea here is that researchers can use semistructured interviews in numerous ways to address numerous research questions. In turn, the quality of such interviews and the resulting analyses can profit from the considerations discussed in this chapter.

6

The Houston Child Study

Some years ago, when I was a faculty member at the University of Houston, I became interested in establishing a partnership with a school in one of the black communities within the city. The idea was to conduct research, yes, but to do so in ways that contributed directly to the school and individual teachers. Thus, with the principal of one specific school, a colleague and I wrote a grant, subsequently funded. The grant provided funds for school materials, summer salaries to the school's principal and science teacher to develop environmental science curriculum, and funds to conduct our psychological research on children's environmental views and values. One of our goals was to use the research findings to enhance the curriculum development. In this chapter, I report on our interviews with the children. In the next chapter, I report on our interviews with the parents. Later, in chapter 12, I bring together both studies to speak briefly about the educational intervention in this specific school, and, more extensively, to place this intervention within a larger perspective on environmental education.

In this study (Kahn and Friedman 1995), we interviewed 72 black children, 24 children (12 males and 12 females) in each of three grade levels: first, third, and fifth (mean ages, 7–5, 9–6, and 11–4). Virtually all the children attending the school were black (99%), and most received the free lunch program (91%). Based on state tests, the majority (60%) of the children were considered low-performing. I should also mention here that in our informal conversations with people in this community, they for the most part preferred to distinguish themselves as black Americans as opposed to African-Americans. Thus, throughout my writing, I follow their preference.

I start with a few excerpts from our interviews to show how human violence was often close at hand in children's environmental reasoning. In this first excerpt, a third-grade girl, Trina, was asked a preliminary question about a bayou—which is more of a southern term for a waterway, like a stream—one of which flowed within a mile of the children's school.

Tell me Trina, do you know what a bayou is? *Yes . . . it's where turtles live and the water is green because it is polluted. People—some people need to, um, some people are nasty. Some people, you know, like some people go down there and pee in the water.* Mm hmm. *Like boys, they don't have nowhere to pee, and drunkers, they'll go do that, too.* Okay. *And sometimes they'll take people down and rape them, and when they finished, they might throw 'em in the water or something.* So, what does it look like? How would you describe it? A bayou? *It's big and long and green and it stinks . . . And turtles live in it.*

Trina clearly knows what a bayou is, and she provides a vivid description of its polluted state ("It's big and long and green and it stinks"). But such environmental knowledge is interwoven with her understandings of how such "natural" states can be used in the inner city: people such as boys and drunks urinate in the bayou, rapes occur alongside it, and bodies are thrown in it.

In the beginning of another interview, a first-grade boy, Billy, was asked what he thinks about in terms of nature and the natural environment. Billy mentions animals, and talks about how much fun he has playing with his dog. The interviewer then probes further:

Are there any other sorts of things that you think about when you think about nature and the natural environment? *My mom.* Your mom? Tell me, why do you think of your mom? *Because one day she went into the store and she left me and my sisters here and she took too long. I thought somebody had raped her.* In the store? Were you left sitting in the car waiting for her? Or you were at school? Or . . . *I was in the car.* Did anything happen to her? *No.* She was alright? *Yes.*

For anyone, let alone a first-grader, that is an awfully quick leap from playing with dogs to rape.

Similarly, consider the perspective of a third-grade girl, Tanya. In response to one of our standard questions ("Are animals an important part of your life? If so, how?"), Tanya says that animals are important because "sometimes they can keep you safe, like dogs, when somebody tries to break in your house." Tanya has good reason for such concern; she quickly moves to the following description:

That's why momma say she going to move out the neighborhood, she say it's too many young people around there . . . too many crack heads, and they do crack and they want to kill somebody. And my uncle, he own the crack, they killed him, but he, he not dead, he's still in the hospital. . . . That's why I always, my momma tell me to stay in the house.

Later, the interviewer asks Tanya if her family ever talks about the natural environment. In the course of responding she says:

That's why we have a little gun in the house, but nobody know where it's at but my grandpa, and he won't tell nobody. He say, it's not to be killing people, it's just to keep this house safe. Well, he not going to kill nobody if they break in, he just going to scare em. So I guess that makes it kind of scary for you sometimes. *Yes, and I always go in the bed with my grandmother 'cause I really scared in the room when my mom 'cause my momma by the window and people be staring up in there, and they be making all the noise, they be shooting up in there, for nothing, just to be shooting.*

Such qualitative data support the commonsense understanding that black Americans face harsh living in urban poverty. But is it also the case that this population has little interest in and affiliation with nature? Our results help answer this question.

A Profile of Children's Environmental Sensibilities and Commitments

For an initial profile of these children's environmental sensibilities and commitments, consider the following results. When asked whether they thought about nature and the natural environment, 96% of the children said yes. They subsequently spoke of animals (59%), plants or trees (54%), various types of pollution (20%), garbage (20%), parks/open spaces (7%), and drugs and human violence (7%). When asked directly, 84% of the children said animals were an important part of their life. Similarly, 87% of the children said plants were an important part of their life, and 70% said parks/open spaces were an important part. Seventy-two percent of the children said they talk about the environment with family members: children reported on conversations that included litter or garbage (47%), air pollution (25%), plants (23%), water pollution (17%), drugs (17%), and animals (13%). Moreover, 74% of the children said they recycle cans and/or bottles, and 25% said they recycle newspapers. When asked if they do anything else that helps the environment, 53% of the children said they pick up litter. Overall, 93% of the children said they or their family do things to help the environment.

Children were also asked to imagine that their entire community threw garbage in their neighborhood bayou. Results showed that the large majority of children believed that harmful effects would result for birds (94%), water (95%), insects (80%), the view (92%), and people along the river (91%). For each of these categories, we also pursued whether children cared that such harm occurred. This additional probe is important, for a factual judgment that harm occurs is not the same as a value judgment about such harm. One might agree, for instance, that a particular action kills an insect without at all caring that that insect is killed. Results showed that children cared insofar as they said it would matter to them if environmental pollution harmed birds (89%), water (91%), insects (77%), the view (93%), and people along the river (83%).

To provide an overall assessment of these children's environmental profile, and to test in one place for effects of age, eleven of the above questions were summed as a single score, reflecting the degree of each child's pro-environmental views and values. The questions included those that pertained to whether the children *thought* about nature, were *aware* of environmental problems, *discussed* environmental issues with their family, *valued* aspects of nature, and *acted* to help the environment. For each question, an affirmative response received a score of one, a negative response a score of zero, and then the scores were summed across the eleven questions. Results showed that out of a possible score of 11 (the most pro-environmental score), first-graders mean score was 7.7, which was statistically less than the score for third-graders at 9.6, and fifth-graders at 9.5.

Children's Moral Obligatory Judgments about Nature: The Case of the Polluted Bayou

At the outset of this study, it was unclear to us what form—obligatory or discretionary—children's moral reasoning would take about polluting a bayou. As discussed in the previous chapter, domain-specific research has found that children usually conceive of prototypic negative moral acts (such as not stealing or not hitting) as morally obligatory. Thus it seemed plausible that children would conceive of not polluting (a negative act) as morally obligatory. On the other hand, the city of Houston discharges its "treated" sewage into its bayous, and by this means the sewage is transported to the ocean. The bayous often smell of pollution, and are not safe for swimming or wading. Garbage is often found along their edge or floating on the water. Thus, with such pollution as a norm, it seemed equally plausible that throwing a small amount of additional garbage into a bayou, while not necessarily desirable, could nonetheless be viewed by the children as permissible. This view might especially take hold in conditions where the pollution occurs in a distant geographical location, where the small amount of increased environmental harm has virtually no direct effect on the children. Indeed a common expression— NIMBY (Not In My Back Yard)—has arisen to characterize the seemingly common phenomenon that people more often object to environmentally degrading acts when those acts occur close to their homes rather than in other parts of their country or the globe.

Results overwhelmingly supported the first proposition. Virtually all of the children (96%) judged the individual act of throwing garbage in their neighborhood bayou as not all right. Children maintained their judgments about not throwing garbage in a bayou even in conditions where local conventions legitimated the practice, for an individual (96%) and for the entire community (94%). Moreover, children maintained that it would similarly not be all right for an individual in a different geographical location to throw garbage in a bayou (96%), even when a different cultural convention legitimated the practice, for an individual (87%) and for the entire community (91%). Basing an assessment of moral obligation on negative evaluations across all six conditions (a stringent assessment, see chap. 4), results showed that 87% of the children viewed polluting a bayou as a violation of a moral obligation. Developmentally, 68% of the first-graders provided such categorical judgments, which was statistically less than for the third-graders at 91%, and for the fifth-graders at 100%. Moreover, for children who judged polluting the bayou as not all right in all six criterion conditions, an analysis was conducted that examined the percentage of children who provided moral justifications (which I will say more about shortly) for their negative evaluations. Results showed that for each question, well over half (depending on the question, from 59% to 80%) of the justifications were moral. In addition, 82% of the children used moral justifications in at least three of their six evaluations. Thus these results support the proposition that moral obligation can be an appropriate construct by which to understand the development of children's moral relationship with nature.

Children's Environmental Justifications

The justification data comprise a rich and complex component of this research, and call for some qualitative presentation to highlight the types of concerns and issues that children brought to bear in their environmental moral reasoning. As shown in table 6.1, two overarching forms of environmental moral reasoning emerged from the data, anthropocentric and biocentric. In anthropocentric reasoning an appeal is made to how effects to the environment affect human beings. For example, consider the following justification for why it is wrong to pollute a bayou:

[It's not all right] *because some people that don't have homes, they go and drink out of the rivers and stuff and they could die because they get all of that dirt and stuff inside of their bodies.*

In this response, the child says that the underlying reason why environmental degradation is wrong lies in the environment's harmful effect on human welfare: that people could die.

A less direct form of anthropocentric reasoning can be seen in aesthetic justifications. Here an appeal is made to ways in which the natural environment can render pleasure to humans in terms of its beauty:

[It is not all right to throw trash in the local bayou because] *the bayou, it should look beautiful. . . . Because, like, if my relatives or something come over, I could take them to the bayou and see, and show them how beautiful it is and clean.*

This reasoning appears to turn centrally on how humans appreciate the aesthetic experience of the natural environment. Thus the child reasons that it is not all right to throw trash in the bayou because a bayou should look beautiful, and that other humans (his relatives) would also like to see a beautiful bayou.

In biocentric reasoning, an appeal is made that the natural environment has a moral standing that is at least partly independent of its value as a human commodity. For example, one form of biocentric reasoning focuses on the intrinsic value of nature, and establishes that value by means of what could be called a naturalistic fallacy in its most literal form:

Because water is what nature made; nature didn't make water to be purple and stuff like that, just one color. When you're dealing with what nature made, you need not destroy it.

This child highlights that what is ("what nature made") ought to remain ("you need not destroy it"). Thus an "ought" is derived from what "is."

A second form of biocentric reasoning focuses on rights for nature. Two ways of establishing such rights appeared. In one way, natural objects (usually animals) are compared directly with humans:

Bears are like humans, they want to live freely. . . . Fishes, they want to live freely, just like we live freely. . . . They have to live in freedom, because they don't like living in an environment where there is much pollution that they die every day.

Thus an animal's desire for freedom ("to live freely") is viewed to be equivalent to that of a human's desire, and because of this direct equivalency animals merit the same moral consideration as do humans. In turn, a second way of establishing rights for nature occurs through establishing indirect compensatory relationships:

Fish need the same respect as we need. Tell me more about this idea of respect. . . . *Fishes don't have the same things we have. But they do the same things. They don't have noses, but they have scales to breathe, and they have mouths like we have mouths. And they have eyes like we have*

Table 6.1
Summary of environmental justification categories

1. Anthropocentric
An appeal to how effects to the environment affect human beings. In other words, the environment is given consideration, but this consideration occurs only because harm to the environment causes harm to people.

A. Personal interests	An appeal to personal interests and projects of self and others, including those that involve recreation or provide fun, enjoyment, or satisfaction (e.g., "[Animals are important to me because] if I go hunting, that's an important part of my life because it'll be fun to me"; "Animals matter to me a little bit because we need more pets and different animals to play with").
B. Aesthetic	An appeal to preservation of the environment for the viewing or experiencing pleasure of humans (e.g., "The bayou should look beautiful because if my relatives come over, I could take them to the bayou and show them how beautiful it is and clean"; "because I'd get to see all the colors of the plants and the beauty of the whole—of the whole natural plants").
C. Welfare	An appeal to the physical, material, and psychological welfare of human beings, including that of agent (e.g., "because if the water is dirty, I might get sick"), of others (e.g., "air pollution goes by and people get sick, it really bothers me because that could be another person's life"), and of society (e.g., "it's wrong to destroy nature because nature will be good for all human kind").
D. Interpersonal condemnation	An appeal to how others would judge the actor(s) negatively for both personal contexts (e.g., "they'd probably lose their friendship with everyone") and publicly (e.g., "harming the environment is wrong because the people in town will get really mad, no one will like these people if they do that").
E. Punishment avoidance	An appeal to punishment or its avoidance (e.g., "because the police might catch her").
F. Influencing others	An appeal to the act's influence on others, with a consequentialist orientation (e.g., "because if a group of people throw theirs in there then a lot more other people will hear about it and they probably will take their trash and throw it in there").

Table 6.1 (continued)

2. Biocentric	
An appeal to a larger ecological community of which humans may be a part.	
A. Intrinsic value of nature	An appeal that nature has value, and the validity of that value is not derived solely from human interests, including is-to-ought appeals (e.g., "if nature made birds, nature does not want to see birds die"; "I think people should care about animals because animals are like part of everyone's life"; "it was here before mankind arrived here").
B. Rights	An appeal that nature has rights or deserves respect, including appeals wherein humans and nature are viewed as essentially similar (e.g., "fishes, they want to live freely, just like we live freely, they have to live in freedom, because they don't like living in an environment where there is much pollution that they die every day"; "animals don't need to be killed either, because they need the same respect that we need"), and set in a compensatory relation (e.g., "Fishes [deserve respect for while they] don't have the same things we have, they do the same things. They don't have noses, but they have scales to breathe, and they have mouths like we have mouths. It's going to be the same, just going to be different").
C. Relational	An appeal to a relationship between humans and nature including those based on psychological rapport (e.g., "animals are important to me because when a person in my family like died, they could come and cheer me up"), personal caretaking (e.g., "I have a dog and he's like my child or something, I take care of him"), and stewardship (e.g., "Those are animals that everyone must take care of, because God put these animals on earth for people to, like for pet stores, to keep and take care of them").

3. Unelaborated harm to nature
An appeal to the welfare of nature, including animals (e.g., air pollution is bad because "the birds and the butterflies, they can't hardly get any air, and it'll probably kill them"), and plants (e.g., "air pollution could kill the flowers and the trees, and the grass and stuff"). No reference is made to whether that concern derives from an anthropocentric or biocentric orientation.

Source: Reprinted from Kahn and Friedman (1995), pp. 1407–1408.

eyes. And they have the same coordinates we have. . . . A coordinate is something like, if you have something different, then I'm going to have something, but it's going to be the same. Just going to be different.

I find this a marvelous passage, as one can feel the constructivist process at work within Arnold. He chooses a word—"coordinate"—that is at once incongruous and precise. It is incongruous because people do not usually use this word in this way. But it is precise because a coordinate can refer to two intersecting index terms that, taken together, refer to a single point. Similarly, Arnold seeks to coordinate two disparate ideas into a unitary position; that is, Arnold appears to draw on a word he encountered in some other context to help him explain that although fish are in some respects not the same as people (they don't have noses like people do), with respect to certain important functions (such as breathing and seeing) they are the same. Thus Arnold moves beyond a reciprocity based on directly perceivable and salient characteristics to establish moral equivalences based on functional properties. In Piagetian terms, Arnold's reasoning could perhaps also be understood as being on the cusp of reversibility, involving the simultaneous coordination of operations (see chapter 3), as he appears not to lose sight of the differences between fish and humans while affirming their functional equivalences.

Now, an important question needs to be asked: How many children like Arnold did we interview? In other words, to what extent did biocentric reasoning emerge as a form of children's environmental moral reasoning?

Toward answering this question, we had systemically probed children's justifications for nine of their evaluations. The first three evaluations (reported above) involved whether animals, plants, and parks/open spaces played an important part in their life. The remaining six questions framed "The Case of the Polluted Bayou." Children's justifications were coded with the categories reported in table 6.1. The resulting justification percentages for each of the nine questions are reported in table 6.2. Averaging across all nine questions, results showed that the majority of justifications were anthropocentric (74%), followed by unelaborated harm to nature (22%), and then biocentric (4%). The "unelaborated" category refers to reasoning that takes account of the wellbeing of nature—"[air pollution is bad because it] could kill the flowers and trees"—

but it remains unclear whether the basis for the judgment refers only to humans (anthropocentric) or includes nature (biocentric). Under the "anthropocentric" category, on average children's reasoning included concerns for human welfare (28%), personal interests (19%) and aesthetics (16%).

From a developmental perspective, several findings, although not statistically significant, were suggestive. Collapsing across all nine questions, results showed that first-graders used the aesthetic category relatively seldom (14%) compared to third-graders (42%) and fifth-graders (44%). In addition, even though biocentric justifications were seldom used, there was some indication of a directional change. Collapsing biocentric justifications across all nine questions (total justifications = 27), the results showed the following usage: first grade (7%), third grade (37%), and fifth grade (56%). Moreover, not one first-grader provided a single intrinsic value or rights-based biocentric justification.

Thus the answer to the question "How many children did we interview who were like Arnold?" is clear: Not many. Still, it is worth calling attention to biocentric forms of reasoning. For one thing, some readers may be surprised that any biocentric reasoning emerged in this population of inner-city children. For another thing, biocentric reasoning may reflect the leading edge of the developmental progression from fifth grade onward (compare Beringer 1994; Nevers, Gebhard, and Billmann-Mahecha 1997). In chapter 8, I will return to this issue with more data in hand and propose that through a developmental process of hierarchical integration, biocentric reasoning may incorporate human-oriented and nature-oriented considerations within a larger mental organizational structure.

A Glimpse of a Possible Mental Disequilibration in the Making

In chapter 3, I offered a structural-developmental account of child development and therein explained the idea of disequilibration. Briefly, it was that early forms of knowledge, although rarely wrong, are incomplete. They are inadequate in addressing competing facts, explanations, and beliefs. When a child (or adult) recognizes such contradictions, and feels their force, a mental imbalance arises, a disequilibration. Since the

Table 6.2
Percentages of environmental justifications by categories

Justification category	Play an important part in your life			Case of the polluted bayou					
				In your neighborhood			In another place, far away		
	Animals	Plants	Parks/open spaces	Individual	Individual: Given the common practice to pollute	Community: Given the common practice to pollute	Individual	Individual: Given the common practice to pollute	Community: Given the common practice to pollute
Anthropocentric									
Personal interests	40	14	86	10	7	8	2	5	3
Aesthetic	4	23	0	19	23	18	19	11	23
Welfare	31	51	11	26	23	26	20	35	25
Interpersonal condem.	0	0	0	0	1	2	6	1	1
Punishment avoidance	0	0	0	7	13	9	9	6	2
Influencing others	0	0	0	2	6	2	16	7	8
Unelaborated	0	0	0	0	0	0	1	0	1
Biocentric:									
Intrinsic value of nature	0	0	0	2	0	1	2	1	1
Rights	4	2	0	2	0	2	2	4	0
Relational	7	4	0	0	0	0	0	0	2
Unelaborated harm to nature	15	6	2	33	27	32	23	30	34

Source: Reprinted from Kahn and Friedman (1995), p. 1411.

disequilibrated state is not a comfortable one, the child seeks to reestablish balance by reorganizing knowledge more comprehensively. Thus, in this way, constructivist theorists posit disequilibration as a central mechanism for development. In this light, I would like to show what such an environmental moral disequilibration might look like by analyzing several more segments of the interview with Arnold, the child quoted above who used the term "coordinate" to establish rights-based reasoning for animals.

Throughout most of the interview, Arnold argues for a principled position that animals deserve the same moral consideration as humans. Equal is equal, he argues, and since people ought not to be killed, so it follows for animals. Yet at times Arnold seems hesitant to follow the full implications of this position. The interviewer, for example, asks Arnold if he ever eats meat. Arnold says no, and talks about how he refrains from eating meat in the school cafeteria. But when the interviewer asks if anyone in his family eats meat, Arnold says, "Only when there's rough times and we really need it." Thus, there is a bit of a tension in Arnold's reasoning: he first categorically objects to killing animals, but then allows for exceptions. Later in the interview, Arnold reiterates his initial position: "Every animals needs respect. And I think they need respect because I love every animal. No matter what life form they're from, no matter how shaped or sized they are." The interviewer responds by posing another potential dilemma:

Do you have the same feeling about mosquitoes as you do about fish? *Well, not really.* [Laughter] Tell me how that's different? *Because mosquitoes they begin to get on your nerves a little bit. And they make little bumps on you. But I don't really like mosquitoes. But it's still wrong to kill 'em though. Because they really need to live freely too, just like every insect, every bear, any kind of type of human, they need to live freely 'cause everybody needs to live freely.*

In places like these, it seems like Arnold comes close to granting—or tries partially to evade—the proposition that sometimes it is morally permissible to kill an animal, be it for food when "there's rough times," or when an animal like a mosquito is particularly annoying. Indeed one can imagine that over time the quantity and force of the contradictions to his categorical judgment against animal killing will increase: perhaps when he encounters the positive role animal research can play in finding cures to devastating human illnesses, or when he learns more about the subsistence hunting of indigenous people, or when he finds himself overwhelmed by mosquitoes one evening, swatting them with a clear-minded intention. Thus in Arnold's development it seems plausible that if and when he directly confronts the validity of even one of these competing claims, his initial conception will prove inadequate. And there it will be, disequilibration. That, in turn, may well set into motion a reorganization of Arnold's conceptual knowledge, perhaps from a concrete, normatively categorical conception of respect for life—"no killing ever"—to one that embeds respect for life within the dynamic changes of ecosystems, and which embraces the troubling truth that at times for something to live, something else must die.

The Urban Context

While the Houston children often demonstrated an anthropocentric and on occasion biocentric orientation toward nature, it should be emphasized that such orientations were constructed in the context of urban living, and this context was often evident in their reasoning. For example, as illustrated at the outset of this chapter, children often wove together environmental descriptions with references to drugs and human violence. Recall Trina, who described a bayou as "big and long and green and it stinks" and noted that "sometimes they'll take people down and rape them, and when they finished, the might throw 'em in the water or something." Or recall that when asked what they thought about in terms of nature, 7% of the children responded with issues pertaining to drugs and human violence; and when asked about what environmental issues they talk about with their families, 17% of the children responded with issues pertaining to drugs and human violence. We had not initially thought to classify reasoning about drugs and human violence as part of environmental reasoning. Yet in hindsight it can surely make sense from the perspective of urban children, where destructive human activity— rather than, say, fresh mountain springs or pristine ecosystems—is part of their immediate environment.

Other times, the urban context shaped children's environmental reasoning in lively if not entertaining ways. Here are two responses from the same child:

[Plants are important] *because we're supposed to keep—take care of all the plants and everything, like people have plant stores and they take care of plants.*

[I care about animals because] *those are animals that everyone must take care of. . . . Because God put the animals on earth for people to, like for pet stores. To keep and take care of them.*

Notice that this child does not say that to take care of plants and animals we should thank God for wilderness, but, in a sense, thank God for plant stores and pet stores. Again, such a focus is not surprising since wild lands are not centrally part of this child's experience.

In this regard, let us turn briefly to one of the children we interviewed who seemed least concerned and connected with nature. She was a first-grade girl, Eboni. Early in the interview, Eboni says that she has a pet cat. The interviewer asks if her cat is important to her. Eboni says, "No. I have other things that's important to me. If I eat or not. Or if anybody in my family is gonna die, because I don't want nobody in my family to die." Here Eboni's reasoning might appear to support a common view based on Maslow's (1975) theory of a hierarchy of needs: that "someone whose needs for food, shelter and physical security are barely met is not likely to spare the energy—physical or emotional—to maintain concern about [environmental issues]" (Hershey and Hill 1977–78, quoted in Mohai 1990, p. 747).

Yet Eboni's stark rejection of animals exists alongside their attraction. Elsewhere in the interview Eboni says she likes dogs, that two of her previous dogs died, and that she wishes she could have another one. Or take the issue of parks and open spaces. On first blush, Eboni rejects them. Eboni never climbs trees. Why? "Cause it's dangerous. Cause if they fall the grass might have glass, and then they fall on their face in the glass, and then they'll cut their nose or eyes and they—they'll be blind." Eboni never walks anymore in the parks. Why? "Because I used to go. Now the people go in there and they be throwing glass and they have guns and stuff and they might shoot me." Indeed Eboni does not even like to play in her backyard. Why? "Nothin' can get me. Like a stranger or something." Is it the case that Eboni has no affiliation with animals, plants, and parks/open spaces? Rather, our data suggests that her economically impoverished and violent urban surroundings have made nature largely inaccessible.

Environmental Generational Amnesia

Houston is one of the most polluted cities in the United States. Within this context, we systematically investigated children's knowledge of three different types of pollution: water pollution, air pollution, and garbage. For each type of pollution, we assessed whether children who understood about the idea of the pollution in general also believed that they directly encountered such pollution in their own city. The findings showed a consistent statistically supported pattern. Sixty percent of the children understood about air pollution in general, but only 36% believed that they encountered air pollution in their own city. Seventy-three percent understood about water pollution in general, but only 28% believed that they encountered water pollution in their own city. Finally, 57% understood about the problem of too much garbage and litter, but only 29% believed that they encountered a problem with garbage and litter in their own city.

How could these children not be aware of their own city's pollution? In the introduction I offered an explanation. I suggested that to understand the idea of pollution one needs to compare existing polluted states to those that are less polluted, and if one's only experience is with a certain amount of pollution, then that amount becomes not pollution but the norm against which more (or less) polluted states are measured. If this explanation is correct, it may capture a phenomenon that can occur any time individuals lack an experiential comparison by which to judge the health and integrity of nature. Consider the following scenario, which informally I have seen played out many times. A family moves to a piece of forested land, say 640 acres, which has already been logged numerous times in the last one or two centuries. The family assesses the land's timber potential. They say: "Hmmm, there should be a way of taking some timber here, and still leave some good trees." Thus they take some timber, and afterward subdivide the land into four 160 acre parcels, keeping one parcel for themselves. A handsome profit. Families from more urban areas now buy each of the remaining 160 acre parcels. In turn, each family now says something like: "Hmmm, there should be a way of taking some timber here, and still leave some good trees." So these families log the land, and afterward subdivide into 40 acre parcels. And the process continues. Notice how relative is the concept of "good." Each logging degrades the land more, but each person assesses the health and integrity of the land relative to the more degraded urban conditions from where they came, and not to the land's condition as it was even a few years before.

Moreover, as I suggested in the introduction, what we discovered in the children we interviewed in the inner city of Houston—and what occurs in the above scenario, if it has merit—might well be the same sort of psychological phenomenon that affects us all from generation to generation. People may take the natural environment they encounter during childhood as the norm against which to measure environmental degradation later in their life. The crux here is that with each ensuing generation, the amount of environmental degradation increases, but each generation takes that amount as the norm, as the nondegraded condition. I call this environmental generational amnesia.

To illustrate this idea further, take a guess at when the following magazine editorial was written:

This [society] is born of an emergency in conservation which admits of no delay. It consists of persons distressed by the exceedingly swift passing of wilderness in a country which recently abounded in the richest and noblest of wilderness forms, the primitive, and who purpose to do all they can to safeguard what is left of it.

In the last decade we have indeed witnessed the swift passing of wilderness in our country, and environmentalists often speak of this problem as one which admits of no delay. The above passage was written, however, in 1935 as the opening to the first issue of the magazine for The Wilderness Society (First Issue 1993, p. 6). Thus environmental problems apparently can be described as equally serious across generations, even while the problems worsen.

Conclusion

The results from this study speak to the importance—personally and morally—of environmental issues in the lives of black children in the inner city. Animals, plants, and parks played an important part in the lives of the majority of the children we interviewed. The children have talked about environmental issues with their family members, and engaged in certain sorts of environmentally helpful behavior such as recycling cans and bottles. They were also aware that water pollution can harm birds, water, insects, and landscape aesthetics. Moreover, harm to each of these environmental constituents (birds, water, insects, and the landscape) mattered to these children. Based on six measures that controlled for magnitude of environmental harm and proximity to harm, most children believed that polluting a bayou was a violation of a moral obligation.

Kellert's survey research (reviewed in chapter 1) on children's attitudes toward nature provides an interesting counterpart to our results. Kellert (1985) found that the moralistic attitude (which includes judgments about right and wrong treatment of animals, with strong opposition to exploitation or cruelty toward animals) increased significantly between the eighth and eleventh grades. Kellert also found that in comparison to white children and children in a rural context, black children in the city revealed less affection and general interest in animals. Our own findings with black children, however, point in a somewhat different direction. Not unlike Kellert, we found an increase in children's moral obligatory reasoning that opposed polluting a bayou, but the shift appeared much earlier: between first graders and fifth graders. Along similar lines, third- and fifth-graders scored higher on our environmental profile than did first-graders. But overall our findings showed that even the first-graders had an environmental moral orientation that was pervasive across a wide range of measures.

Children's environmental moral reasoning largely focused on anthropocentric considerations (e.g., that nature ought to be protected in order to protect human welfare). With much less frequency children focused on biocentric considerations (e.g., that nature has intrinsic value or rights). Environmental philosophers often argue for the superiority of biocentrism over anthropocentrism (Callicott 1985; Regan 1983; Rolston 1981, 1994, 1995; Taylor 1986), but even if they are right, I want to stress that anthropocentric reasoning should not be discounted. Indeed environmental philosophers have occasionally argued that anthropocentrism represents the only valid basis by which to ground an environmental ethic. Baxter (1986), for example, argues that people need only protect nature insofar as such efforts benefit human interests: "Damage to penguins, or sugar pines, or geological marvels is, without more, simply irrelevant" (p. 215). In this view, people may decide to protect penguins "because people enjoy seeing them walk about rocks" (p. 215). A person may "even decline to resist an advancing polar bear on the ground that the bear's appetite is more important than those portions of himself that the bear may choose to eat" (p. 216). But it is still a person who does the valuing, not nature.

Similarly, Wilson (1984) writes that the "only way to make a conservation ethic work is to ground it in ultimately selfish reasoning" (p. 131). Wilson does so on many occasions. For example, Wilson (1992) writes that it is fashionable in some quarters to wave aside the importance of biological diversity,

> forgetting that an obscure moth from Latin America saved Australia's pastureland from overgrowth by cactus, that the rosy periwinkle provided the cure for Hodgkin's disease and childhood lymphocytic leukemia, that the bark of the Pacific yew offers hope for victims of ovarian and breast cancer . . . and so on down a roster already grown long and illustrious despite the limited research addressed to it. (p. 345)

I am not saying that the anthropocentric reasoning of the Houston children is identical to that of Wilson's. For while Wilson (1984) is after "selfish reasoning . . . the premises must be of a new and more potent kind" (p. 131), the kind he articulates through the construct of biophilia, discussed in chapter 1. But I am saying that the Houston children articulated forms of anthropocentric reasoning that many would argue are not only philosophically valid, but practically speaking—in terms of effecting societal change—an imperative.

Finally, studying children's relationship with nature in this urban context provides a unique perspective on the biophilia hypothesis. Recall from chapter 1 that Wilson (1984) suggests that one excellent way to investigate biophilia is through studying the landscapes that wealthy individuals inhabit when free from their work (e.g., where they go to vacation, where they build summer residences, and so forth). His reasoning is that people who are largely free from economic and time constraints would most exhibit "natural"—genetically based—inclinations, and inhabit landscapes that from an evolutionary standpoint contribute to survival and reproductive success (e.g., water access, bluff tops, and savannalike landscapes). It also seems the case, however, that further support for biophilia could come by studying people who are not wealthy but, on the contrary, extremely poor and living in an inner city. For if an affiliation with nature can be shown to exist even in those people most encumbered by economic and urban constraints, then that would speak to pervasive and deeply abiding biophilic characteristics. This current study provides such support insofar as it appears that the serious constraints of living in an economically impoverished urban community cannot easily squelch these children's diverse and rich appreciation for nature, and moral responsiveness to its preservation.

7

The Houston Parent Study

Black communities in the United States are disproportionately subjected to large amounts of environmental hazards and pollution (Bullard 1990; Gaylord and Bell 1995; Wenz 1995; Westra 1995). However, little is known about this group's environmental concerns, understandings, and values (Bullard 1987; Mohai 1990). Thus, in this second study (Kahn and Friedman 1998) we interviewed 24 black parents who had at least one child enrolled in the school discussed in the Houston child study. We pursued four overarching questions: How do parents value the importance of nature for their family? What environmental problems are parents concerned about? What types of environmentally related behaviors do parents participate in with their family? What are parents' views toward environmental education for their children? Our goal was to help characterize and give voice to black parents' perspectives on nature and environmental education.

Before moving forward, I should mention that we defined a "parent" as the child's primary caretaker, which sometimes was the child's grandmother (8%) or other guardian (8%). Of the parents we interviewed, 23 were female and 1 was male. Two reasons can help explain this disparity. For one thing, the interviews occurred during school hours (on the school grounds), and fathers were more likely than mothers to be employed during that time. It was also the case that men, as parents, resided in a small percentage (23%) of the households that participated.

The Importance of Nature

Eighty-six percent of the parents said animals played an important part in the lives of their family. Often, on this topic, parents spoke favorably of pets ("we're crazy about animals, pets"). It was also not unusual for parents to describe how their children interacted with a diverse range of smaller animal life close at hand:

My grandson picks up all kinds of little animal things and some of them, I don't even know what they are myself, but he brings them in and gets a jar.

My kindergarten daughter, she might see something that looks injured and, um, she saw a worm. She doesn't pick up these black ones or brown ones because they sting. So this one was a yellow one and she said he was hungry. So she picked him up and took him over to a leaf and put him on it. You know, they do those type things.

Eighty-six percent of the parents also said plants played an important part in the lives of their family, and 95% said parks and open spaces did. Often parents spoke with enthusiasm of these aspects of nature ("we love plants"; "my children love to run in the park").

Parents described at least two types of problems that made it difficult to interact well with their natural surroundings. One problem involved pollution:

[Where I live] they have a lot of backup sewage and stuff. And my children can't play in the backyard because it's just nasty. And right now before I left they was over there trying to stop it. It'll be right back. Back up probably next week some time.

A second problem involved social violence:

Because kids don't have no break around here because all they could do is stay in the house. They couldn't really go outside and sit on the porch, 'cause somebody may shoot 'em or something. 'Cause it was just that bad around here. But now, it's just a little better, 'cause I guess the police had really got on their jobs.

Such concerns echo those of the children, like Eboni, reported in the Houston child study. In other words, it is not so much that parents and their children wanted to avoid nature (even in terms of just playing in the backyard or sitting outside), but that the noxious pollution and potential for violence within their community made such experiences difficult.

All of the parents (100%) said it was important for people to live in harmony with nature. Parents' conceptions of what it means to live in harmony with nature were coded with the categories presented in table 7.1. Results showed the following percentages of the total number of conceptions offered (multiple conceptions were coded): acting upon nature (39%), experiencing nature (18%), being in the right state of mind with nature (13%), being in balance with nature (13%), and respecting nature (16%).

Although we did not pursue this study developmentally, it is possible that the latter forms of reasoning, particularly those that appeal to balance with nature and respect for nature, emerge through the hierarchical integration of earlier forms (a process discussed in chapter 3). For example, as presented in table 7.1, consider a typical appeal to balance with nature: "[Harmony means] you're balanced out with nature, to where you're not working against it, like we can't exist without plants, and without us they can't exist." Granted, this parent asserts facts that seem incomplete. For although it is true that many domestic plants may not exist for long without human support, most other plants do very well indeed. But, regardless, each reason by itself—whether people help plants or plants help people—is act-centered. In turn, when both acts are placed within a reciprocal relation, the concrete reasoning would appear to undergo a qualitative change which thereby allows for the concept of balance with nature to emerge. I will come back to this idea in the next chapter, where I provide developmental data on children's conceptions of harmony.

Across five questions, we coded parents' environmental justifications with categories like those used in the Houston child study (table 6.1). Three questions focused on why parents believed plants, animals, and parks were or were not an important part in their lives, the fourth on why parents proposed particular types of environmental curriculum, and the fifth on why parents believed it was important to live in harmony with nature. Summing across the questions, results showed the following percentages of the total number of justifications offered (multiple justifications were coded): personal interests (24%), aesthetics (16%), welfare (39%), relational (12%), and biocentric (10%).

One of the main differences between these categories and those reported in the Houston child study is that we no longer classified relational reasoning as biocentric. Our reason was that in looking at more data we

Table 7.1
Parents' conceptions for living in harmony with nature—summary of categories

Acting upon	Conception based on doing something to or for nature, including *positive acts* ("to live in harmony with nature means to help the environment"; "[harmony means] planting more trees"; and *negative acts* ("not polluting the air"; "don't be shooting at the birds").
Experiencing	Conception based on experiencing or interacting with nature ("[harmony means] being out in nature"; "just going to a river or lake or something and just sitting there, absorbing all of the fresh air, the outside").
State of mind	Conception based on experiencing a particular state of mind or feeling ("[harmony means] to enjoy the outside"; "to live happily together as one big happy family").
Balance with nature	Conception based on being in balance with nature ("[harmony means] you're balanced out with nature, to where you're not working against it, like we can't exist without plants and without us, they can't exist"; "working together, because everybody [including a person, ant, or mouse] has a job to do or a place").
Respect for nature	Conception based on respecting nature, including such concepts based on reciprocity ("[harmony means] I'm going to respect the bee, if he respects me") and perspective taking ("to put themselves in the animals' shoes, could they live in that environment with all the air pollution living outdoors").

were not satisfied that relational reasoning sufficiently grants moral standing to nature. Here is an example of what I mean. One parent said:

[Animals are important in my family's life] *'cause I love children, and animals to me are just an emulation of a child. . . . And you bond with them. Your love for them becomes almost at times, of love that you can have for human beings because of the way that they in return show you their affection.*

Such reasoning clearly establishes a psychological relationship between humans and animals ("you bond with them"). But the relationship largely appears anthropocentric, cast in human terms ("your love for them becomes almost at times, of love that you can have for human beings") and for human benefit ("they in return show you their affection"). Although I still find it a hard call, the advantage of not classifying such reasoning as biocentric is that it reserves that term for unequivocal cases. That in turn provides a conservative assessment of the amount of biocentric reasoning in any population.

Environmental Problems

All of the parents (100%) were aware of at least some environmental problems, including air pollution (75%), water pollution (71%), and garbage (67%). Most of the parents (91%) believed they were directly affected by one or more environmental problems, such as air pollution (54%), garbage (42%), and water pollution (21%). Their knowledge was often direct:

I'd say about the third week, I have gotten up early in the morning and walk outside and the pollution smell, like, really bad. Sometimes, I'll tell you what, it seems like sometimes you come by and it smells like a cesspool, but it's really not. You can smell that chemical and where it's coming from, but then you have to go back 'cause sometimes it be real strong. . . . And smells strong, sometimes it smells terrible.

[The air] stinks, 'cause I laid up in the bed the other night. Kept smelling something, knew it wasn't in my house 'cause I try to keep everything clean. Went to the window and it almost knocked me out. The scent was coming from outdoors into the inside and I didn't know where it was coming from. . . . Now, who'd want to walk around smelling that all the time?

These findings parallel those from national surveys where it is routinely found that over 90% of respondents express concern about environmental pollution (Kempton, Boster, and Hartley 1996).

Since these parents were aware of various forms of environmental problems, and experienced many directly, we also asked them to evaluate two overall strategies for solving them. Specifically, we asked the following question: "Some people say there are two ways to solve environmental problems. One way is to decrease our consumer needs and decrease our use of technology to control nature. The other way is to push ahead with developing new technology since it is believed that technology will be able to solve the environmental problems. What do you think? Do you favor one way more than another? Why?"

The results showed that 67% of the parents favored conservation over technological solutions, while 33% favored technological solutions over conservation. Here are two examples of each type of solution:

Nature is natural and with all this high tech we have going on now, it's not really guaranteed. You know what I'm saying? But we can always depend on nature 'cause we come from nature. Nothing takes the place of nature. (Conservation Solution)

I don't feel we should do the technology because in a sense they always have new ways of doing things, and then when they get through with that project there's something else they didn't remember to do. (Conservation Solution)

Going back to nature, that's not going to get it, is it? I mean, things have changed and you got to change with it. You know, nature's fine, but I think going on with new technology would be the better way. Like when a big oil spill or something happens, you don't have the technology to clean it up, and going back to nature, it's just not going to work. (Technological Solution)

[The problems have been going on] *so long and it's still going on. So I think they need to come up with some new development and technology.* (Technological Solution)

These responses appear to fit into the larger societal discourse on ways in which technology can both enhance and degrade human welfare and the human connection with the natural world (Kohak 1984; Rothenberg 1993; Strong 1995).

Environmental Practices

The majority of parents (93%) said they did things to help the environment. Activities included recycling cans and bottles (70%), picking up litter (43%), recycling newspapers (26%), recycling other items (22%), and reusing materials (4%). Some of the parents who recycled cans and bottles did so by giving their recyclables to other people who went from house to house requesting such items. These other people, in turn, took the recyclables to a facility (which was located outside their community) and collected the money. Parents often gave their recyclables away in this manner because they did not own a car and thus lacked the means to recycle the materials themselves. This lack of transportation sometimes affected parents' environmental practices in other ways, as well. For example, when parents said they did not often go to parks, the reasons sometimes stemmed from not having a car to get to the nicer and safer parks outside their community. Similarly, a lack of transportation sometimes limited the environmental educational opportunities between parent and child: "A lot of parents don't take their kids a lot of places where they can understand [things about nature] . . . because a lot of parents don't have no transportation."

The majority (88%) of the parents had conversations with their children about environmental issues, such as water pollution (18%), garbage (15%), harm to plants (15%), air pollution (12%), harm to animals (9%), recycling (9%), and chemicals in food (3%). These family conversations were started in a variety of ways, based, for example, on observing and interacting with nature directly (47%), TV and movies (47%), school discussions (27%), and newspapers or other media (7%). These conversations were often poignant:

Yesterday, as my son and I were walking to the store and we were walking down Alabama [Street], and for some reason I think they're getting ready to widen the street. And it's a section of Alabama that I thought was so beautiful because of the trees, and they've cut down all the trees. And you know it hurts me every time I walk that way, and I hadn't realized that my son had paid attention to it, too. So he asked me, he said, "Mama, why are these, why have they cut down all the trees?" And then he asked me, "Well, if they cut down all the trees everywhere, would that have an effect on how we breathe?"

The water we drink just comes out of the faucet and sometimes he'll say something like, "This water doesn't look right." You know, it could have something in it that could be detrimental to us. [My son asks], "Could it hurt me? How do we know what's in this water?" And to some of his questions I have no answer because, I mean, I cannot tell him what's in the water 'cause I don't know. I wonder some things myself.

Such conversations point to an appreciation for nature (of trees), environmental concerns that arise through direct experience of environmental degradation (the cutting of trees and water pollution), and perhaps some sense of powerlessness in not being able to preserve what exists of their community's natural beauty and in not knowing about their environment's safety.

Parents encountered at least two problems that appeared to limit their pro-environmental behavior. One was that once a parent might solve a local environmental problem, one or more other residents in the community would create it again:

There is a lot of trash, and it just makes the area look ugly. And you can clean up and then go to bed and wake up—the people that walking up and down the street, you know?—wake up in the morning and it's right back throwed in your yard.

Right open daytime, they just come and take [the flowers I plant]. *Yeah, because I try to keep some of 'em outside, you know. I try to make it look pretty out there and they just, you know, they'll come and take 'em.*

A second was that solutions to some environmental problems depended on the cooperation of larger numbers of people:

I guess it makes it hard because, you know, whereas maybe I might be doing it, my—my neighbor next door might not be doing it.

Well, to tell the truth, the little bit that we do do, it might be helping out a little bit, but it's not too much one single family can do . . . maybe things wouldn't be exactly the way it is right now, but one family's not going to make that much difference.

Thus—perhaps like for environmentally concerned individuals everywhere—a sense of futility at times emerged in the voices of these parents.

Environmental Education

Research has shown that if education is to succeed better in black communities it will likely depend in part on support from the home (Ogbu 1977, 1990, 1993; Parental Role 1994; Solomon 1992; Winters 1993). To what extent did these parents support environmental education? We pursued this question. On a scale of 1–10 (with 1 the least important and 10 the most important), we asked parents to rank the importance of drug education for their children. Results showed a mean rank of 8.5. On the same scale, we asked parents to rank the importance of environmental education for their children. Results showed a mean rank of 8.7. Statistical tests showed no difference between parents' rankings for the importance of drug education versus environmental education. In comparison to environmental education, 57% of the parents ranked drug education as equally important, 29% as more important, and 14% as less important. Of parents who equated the importance of drug and environmental education, their reasoning often focused on the physical ramifications of both problems:

With the drugs, we're nothing. Without the environment, we're nothing. And drugs is something I see every day. There are dealers across the street from me. So I see this every day and it's just killing us. I mean, it really is killing us, and with the drugs we're not going to have any youth. . . . With the drugs you're not going to have a future and without any environment we're not going to have a future.

Well let's put it like this here. If you don't take care of one [drugs], *it's going to kill you. If you don't take care of the other* [the environment], *it's going to kill you.*

We also asked parents what they thought would be important for their children to learn about nature and to include in their children's school curriculum. Based on the total number of responses (multiple responses were coded), parents suggested a focus on littering/garbage (16%), air pollution (14%), spiritual aspects of nature (12%), plants (12%), animals (12%), water pollution (6%), drugs and human violence (4%), technology (4%), recycling (4%), and nature walks (4%). All of the parents (100%) favored environmental education that coordinated school curriculum with at-home activities.

An Historical Perspective

The political activist Eldredge Cleaver (1969) once wrote that "black people learned to hate the land . . . [and] have come to measure their own value according to the number of degrees they are away from the soil" (pp. 57–58). Others have similarly advanced the proposition that certain conditions—such as a history of slavery—have denied blacks the opportunity to develop appreciative attitudes toward nature (for an overview of the literature, see Taylor 1989; Mohai 1990). Such a perception may have been fueled historically by the large migration of black Americans until the 1970s from the rural South to the cities of the North and West. Yet, in contrast, many black writers—from Booker T. Washington to James Baldwin to Huey Newton to Toni Morrison and bell hooks—

have written of the enduring place of land and nature in the psyche of black Americans. For example, bell hooks (1996) writes:

Living in modern society, without a sense of history, it has been easy for folks to forget that black people were first and foremost a people of the land, farmers. . . . Living close to nature, black folks were able to cultivate a spirit of wonder and reverence for life. Growing food to sustain life and flowers to please the soul, they were able to make a connection with the earth that was ongoing and life affirming. They were witnesses to beauty. (p. 21)

According to hooks, "generations of black folks who migrated north to escape life in the South [have] returned down home in search of a spiritual nourishment, a healing that was fundamentally connected to reaffirming one's connection to nature. . . . " (p. 22). Such desires for a stronger connection to nature may have contributed to the reverse migration that Stack (1996) and others have begun to document. By 1990 the South had regained a half-million black Americans, both in the Southern cities and countryside.

Something of this history, and the black American's connection to the land, emerged from our own data. Some parents spoke of growing up in the country and their desire to impart some of that way of life to their children:

I was born in Mississippi and I spent a great part of my life on a farm. And on the farm we had corn and tomatoes and okra and stuff like that, and we had a couple horses, and before my father passed away my baby was able to spend about a year or so there. And he just loved that. As a matter of fact, if possible he would go back to that kind of life.

I'm from the country. So I want them to really learn all the outside things, you know, that I enjoyed as a child coming up. You know, fresh air, and I enjoyed fishin' and stuff like that, so I want them to be able to, you know, go out there and enjoy nature.

Other times, parents' connection to the countryside was through their own parents and grandparents:

When I was a little kid, my grandmother lived out in the country. She said she liked it out there 'cause it was quiet and peaceful and you can breathe—the air out there was clean and fresh.

Such connections to a rural past—although largely unexplored in this study—may further explain these parents' particular receptivity for environmental education for their children.

A Methodological Qualification

At the outset of this study, the principal of the school objected to our proposal to solicit parents randomly from the school population to participate. He said that such a solicitation would be too burdensome on parents, and in any case largely ineffective given that most parents did not communicate with the school through such written documents. Instead, the principal targeted certain parents to recruit. From our informal discussions with the principal it appeared that he (reasonably) chose to solicit parents that had been somewhat active in the school or in their child's education. But how much did this "principal-solicited" population of parents differ from a representative population within this community?

One answer is that our results may represent a higher bound in terms of this community's overall environmental orientation. If so, then our results would still show an important characteristic of this community and perhaps most if not all urban black communities in the United States: within these communities live at least some environmentally oriented parents. Moreover, our results would help flesh out the depth and complexity of such parents' orientations in terms of their environmental commitments, values, and reasoning.

Alternatively, another answer is possible—one which emerges from recent comparative research between black and white environmentalism. Mohai (1990) reanalyzed a large set of survey data conducted by Louis Harris, Inc. (Fischer et al. 1980) and found higher levels of black environmentalism than was often reported in the research of the 1970s (Crenson 1971; Hershey and Hill 1977–78; Hohm 1976; Kreger 1973; Mitchell 1979; Ostheimer and Ritt 1976). Moreover, in Mohai's analyses there were no statistical differences in environmental concern when blacks and whites were compared as a whole, and few differences when blacks and whites were compared by socioeconomic categories (p. 754; see also, Bryant and Mohai 1992; Bullard 1990; Mohai and Bryant 1996; Mohai and Twight 1987). Where Mohai found statistical differences, blacks scored higher than whites on environmental concern in four out of the seven cases. Thus to the extent our results are of a piece with white environmentalism, and indeed of national samples (Kempton, Boster, and

Hartley 1996), the explanation may lie not in the potential bias of our sample, but that in the United States there is less difference between black and non-black communities than some individuals might perceive.

Conclusion

The black communities in Houston have remained—in the words of Robert Bullard (1987)—largely invisible to politicians, researchers, and environmentalists alike. Through interviews with parents, this study helps to make visible one black Houston community's perspective on nature and environmental education.

In summary, parents spoke of their commitment to environmental issues and enjoyment of nature. Animals, plants, and parks played an important part in the lives of these parents and their families. Parents were aware of the negative effects of environmental problems such as air pollution, water pollution, and garbage. Their knowledge was often direct, visceral: the air would often "smell like a cesspool," and sewage would often back up and be "just nasty." Parents talked about such issues with their children. In response to environmental problems, parents more often favored conservation solutions over technological solutions. Parents acted to help the environment, often by recycling. Parents were also committed to environmental education for their children. In terms of their environmental reasoning, parents drew most often on anthropocentric considerations, including personal interests, human welfare, and aesthetics. Yet biocentric reasoning—where nature itself is given moral standing—was not entirely absent, comprising 10% of all justifications. Moreover, more than one quarter of parents' conceptions of living in harmony with nature involved the biocentric orientations of being in balance with nature or respecting nature.

Finally, it is important to note that our general approach in this study shifts the ground slightly of the research enterprise. Instead of only comparing the black commitment to nature and environmental issues to another racially-based population, it is also important to ask how the black relationship with nature is to be understood within the context of their social and physical environment. How can we build on such relationships to foster environmental education and a healthier and more life-affirming connection with the natural world? These questions support the recent call from Gates and West (1996) when they write: "We [the black communities] need something we don't yet have: a way of speaking about black poverty that doesn't falsify the reality of black advancement; a way of speaking about black advancement that doesn't distort the enduring realities of black poverty" (p. B7). Indeed, through our interviews, black parents gave voice to both realities. They described the harsh living of urban poverty, from drive-by shootings to drug dealers living next door, while articulating, sometimes eloquently, their environmental awareness, values, and sensibilities, and guarded hopefulness for their children's future.

8

The Prince William Sound Study

On March 24, 1989, the *Exxon Valdez* supertanker ran aground in Prince William Sound, Alaska. The tanker's hull ruptured, and nearly 11 million gallons of crude oil spilled into the Sound. This oil spill is the largest one to occur in North America and "the most destructive single event of oil pollution in North American history" (Keeble 1993). Although the full effects are still hotly contested, it appears that this oil spill killed thousands of marine mammals and more than a quarter-million birds, deposited over one million gallons of oil on beaches and shoreline, harmed the ecosystem of the Sound for at least decades, harmed the subsistence livelihoods of Native Americans, led to potentially long-term psychological disorders of residents within local communities, and resulted in many billions of dollars of economic damage (Gilardi 1994; Holloway 1991; Keeble 1993; Matthews 1993; Pain 1993; Palinkas et al. 1993).

Environmental disasters of this magnitude capture the attention of millions of people. They also shape social discourse and environmental practice for years to come. However, little is known about how such disasters impact the minds and hearts of children at large. Thus, during the spring of 1990, my students and I interviewed sixty children about their understandings and values related to this oil spill (Kahn 1997b). There were twenty children in each of three grade levels: second (mean age approximately 8–6, 10 females and 10 males), fifth (mean age, 11–7, 12 females and 8 males) and eighth (mean age, 14–5, 14 females and 6 males). The children were selected from two public schools in Houston, Texas, and represented diverse ethnic backgrounds and levels of economic standing. Throughout this study, we sought to understand these children's responses to a real-life environmental disaster. We also sought to provide an increasingly sophisticated account of the development of children's moral environmental reasoning.

Children's Understandings of the Harm Caused by the Oil Spill

Kick a dog. It might howl and then run, and then shy away when you next approach. Kick a rock. By most accounts it does not utter a sound, feel, or otherwise respond in a sentient fashion. These two cases highlight a basic distinction of categories in the natural world, between nonsentient and sentient nature. Thus throughout this study we asked questions about the effects of the oil spill on shoreline (nonsentient nature) and marine life (sentient nature).

One initial line of investigation focused on whether children believed the oil spill harmed marine life and shoreline. We also asked questions concerning harm to humans in terms of fishermen, recreational users, and the Exxon Oil Company. Results showed that the large majority of the children believed the oil spill harmed the local Alaskan shoreline (90%), marine life (100%), fishermen (98%), recreational users (100%), and the oil company (94%). Then for marine life and shoreline, we also assessed (as we did in the Houston child study) children's value judgments about whether the harm matters. Of the children who had recognized harm, results showed that the majority said it would matter to them that the harm occurred to the shoreline (86%) and marine life (95%).

This distinction between whether harm occurs to sentient and nonsentient nature forms part of a larger typology of environmental harm. There are at least four additional distinctions of importance, which could be explored systematically in future research. One is whether the environmental harm is minor (littering) or massive (the Prince William Sound oil spill). A second is whether the environmental harm arises incrementally or immediately. Spilling oil in a body of water, for example, can be compared to ingesting lead into the human body: a minuscule amount makes little difference, but as the concentration level increases, toxicity of the larger system results. A third is whether the harm arises due to nonhuman forces (hurricanes) or human activity (transporting oil). A fourth is whether the environmental harm is caused accidentally or intentionally. An oil spill usually occurs accidentally; yet note that under the leadership of Saddam Hussein during the Persian Gulf War, Iraq intentionally dumped 250 million gallons of oil into the Persian Gulf.

Children's Moral Obligatory Reasoning about the Oil Spill

Given that children recognized that the oil spill caused harm to both nonsentient and sentient nature, and that the children cared about that harm, we also pursued whether children believed that humans violated a moral obligation by means of the oil spill's harmful consequences. As discussed in chapter 4, we assessed moral obligation using three criteria: prescriptivity, rule contingency, and generalizability. Results showed that the majority of children said it was not all right that the oil spill harmed the shoreline (98%) and marine life (96%), that such harm would not be all right even if a law allowed for it (97% for shoreline and 98% for marine life), and that it would not be all right even if it happened in a far off place where people thought the act would be all right (89% for shoreline and 82% for marine life). Based on a conjunction of all three criterion judgments, the results showed that the majority of children viewed the act of polluting the shoreline (86%) and marine life (75%) as a violation of a moral obligation.

The moral quality of children's obligatory judgments are underscored by those justifications that appealed to welfare, justice, intrinsic value of nature, and unelaborated harm to nature. For these justifications turn on considerations of justice, welfare, and virtue—issues, as discussed in chapter 4, that in moral philosophy traditionally come under the purview of morality. Accordingly, for children who provided negative evaluations on all three criterion questions, we examined the percentage of children who provided moral justifications. For shoreline and marine life, the results showed that 96% and 100%, respectively, of the children provided a moral justification for a least one of their three evaluations (73% and 86% for two of the three; and 50% and 56% for three of the three).

Developmentally, however, second-graders less often generalized their prescriptive judgments for shoreline and marine life than the fifth-graders and eighth-graders. A similar developmental pattern occurred in the Houston child study, and has appeared in other studies using prototypical (nonenvironmental) moral stimuli involving issues of fairness and welfare (Davidson, Turiel, and Black 1983; Tisak and Turiel 1988). Domain theorists have pointed out that this lack of generalizability usually occurs only with young children, particularly those who are largely unfamiliar with the event under discussion (Smetana 1995). That may be the case in this study involving an oil spill that occurred thousands of miles from where these children lived. But it was not the case in the Houston child study, where the stimuli involved throwing garbage in the children's local bayou.

How, then, should we interpret these developmental findings? As discussed in chapter 4, from a Kohlbergian perspective one might believe that our findings provide evidence for children's lack of differentiation between moral and nonmoral concepts. Other results from our studies, however, show differentiations such as those based on the criterion of rule-contingency and patterns of justification use. Thus I am more inclined toward a domain interpretation that the differentiations exist but perhaps are not yet fully consolidated in early childhood. I am also inclined here to follow Glassman and Zan (1995) and recognize that this phenomenon has yet to be well understood.

Children's Moral and Ecological Justifications

Children's justifications for their evaluative judgments were elicited for ten of their above evaluations. The resulting justifications were coded with the categories reported in table 8.1. The quantitative results are reported in table 8.2, broken down by each of the ten questions.

Recall from the Houston child study that we asked the question: How many children—like Arnold—used biocentric reasoning? The answer was: not many. Similarly, here we assessed quantitatively children's use of anthropocentric and biocentric reasoning, and then tested for developmental effects. The analysis proceeded as follows. First, the individual justification categories were collapsed into the most overarching categories: harm to nature, anthropocentric, and biocentric. Then the mean proportionate use of each category was calculated by grade across all ten questions. Results showed the following pattern: harm to nature: 26% (second), 35% (fifth), and 22% (eighth); anthropocentric: 42% (second),

64% (fifth), and 58% (eighth); and biocentric: 8% (second), 26% (fifth), and 20% (eighth). Statistical results showed that in comparison to the second-graders, the fifth-graders and eighth-graders used a greater percentage of anthropocentric and biocentric reasoning. I shall come back to the implications of these findings in the following section on the development of biocentric reasoning.

As discussed earlier in this chapter, based on their evaluations children seldom distinguished between nonsentient and sentient aspects of nature. However, differences emerged based on children's justifications. Specifically, the four shoreline questions (of value, prescriptivity, rule-contingency, and generalizability) were combined, and their mean proportionate justification use calculated. The same procedure was used with the four marine life questions. Statistical results showed that "harm to nature" was more often used for the shoreline (mean use, 32%) than marine life (22%) stimuli. Similarly, "anthropocentric" was more often used for the shoreline (70%) than marine life (50%) stimuli. In contrast, "biocentric" was more often used for the marine life (26%) than shoreline (4%) stimuli. Thus our results suggest that in comparison to nonsentient aspects of nature, the sentient aspects pull more toward biocentric responses.

Let me here also point to three particularly interesting extensions of our justification coding system, in terms of anthropocentric welfare reasoning, natural processes, and telos of nature.

Anthropocentric Welfare Reasoning

The Houston child study had delineated anthropocentric environmental welfare reasoning based on considerations of physical, material, and psychological harm. In the current study, two other types of harm justifications emerged. One was harm to individuals within a larger systemic social context, including political and economic systems. For example, one child said: "It's not all right because we're killing the fish, we're killing the economic process of that state." Here there is an ability to recognize that harm to nature can harm the prosperity of people living within statewide economic systems. A second type of harm focused not on human systems but ecological systems. For example, one child said: "Because, like, if [the oil spill] was in Australia or somewhere it would

Table 8.1
Summary of environmental justification categories

Category and type of response	Summary description
Anthropocentric	An appeal to how effects to the environment affect human beings
Personal interests	An appeal to personal interest and projects of self and others, including those that involve recreation or provide fun, enjoyment, or satisfaction ("[it matters] because they can't have their favorite food or do their hobby . . .")
Welfare	An appeal to the physical, material, and psychological welfare of human beings, including that of self, of other individuals, and of individuals within a larger systemic social context or ecological context ("it wouldn't be all right because like if it was in Australia or somewhere, it would eventually pass on to us and mess up, because we're all the world, you know, and it's going to eventually come to us")
Educative	An appeal to the potential for humans to learn from nature ("because if we lost the endangered species of the fish in the oil spill, we won't be able to learn physically and mentally from them")
Justice	An appeal to fairness or the rights of other humans, including a focus on locus of responsibility and unjustified harm (it's not all right because everyone has a right to work)
Aesthetics	An appeal to preserving the environment for the viewing or experiencing pleasure of humans ("because we might not see them beautiful fish no more if they were killed")
Biocentric	An appeal to a larger ecological community of which humans may be a part
Intrinsic value of nature	An appeal that nature has value that is derived not only from human interest, including a focus on biological life, natural processes, establishing value equivalencies between other life forms and humans ("because if it was human lives, then it would still be the same thing, it wouldn't be all right because it's lives"), or a teleos of nature ("without animals, the world is like incomplete, it's like a paper that's not finished")
Justice	An appeal that nature has rights or deserves respect or fair treatment ("it's not all right because I think every creature, people, or thing or whatever has a right to live"), including appeals to unjustified harm ("it's not all right that the oil spill killed many animals because I don't think it was their fault"), or established by means of a direct relation between

Table 8.1 (continued)

	humans and nature ("because I think fish and animals have a right to live just like we do, and it's not fair to have killed them this way"), a compensatory relation ("just because of their appearance and they can't talk, they're animals, and I don't think that's right, they could be people if they could talk, a form of people, well not human beings but something like it, just a degree of level and that's it, that's the only difference") a conditional relation ("it's not all right because they're dead, living things just like we are, you wouldn't want anybody to kill you like that"), or a hypothetical perspective-taking relation ("you put yourself in the animal's position and you wouldn't like that, and so if you just kind of trade places and think about it, and everyone would think it wasn't right")
Unelaborated harm to nature	Although no reference is made to whether appeals for nature derive from an anthropocentric or biocentric orientation, such appeals include a focus on animals, nonliving parts of nature, food chains, and ecosystems ("it wouldn't be all right because if the animals die, the land wouldn't be fertilized to grow plants, and animals need plants to eat, and when the animals give out carbon dioxide, plants suck it in to make oxygen, and the animals need oxygen to live")

Source: Reprinted from Kahn (1997a), p. 1093.

eventually pass on to us and mess up, because we're all the world, you know, and it's going to eventually come to us." Notice that this reasoning is still anthropocentric because it focuses on how harm to nature harms human welfare. But there is the understanding that such harm can occur because of interconnected ecological systems ("because we're all the world, you know").

Natural Processes
Biologists have had some difficulties assessing the amount of long-term environmental harm caused by the Prince William Sound oil spill. One reason is that no fixed standard exists by which to compare later changes. For example, researchers supported by the Exxon corporation point out that in 1993 the Sound experienced a heat wave and that mussels and other intertidal life forms in the Sound do poorly in warmer tempera-

Table 8.2
Percentages of environmental justifications by questions and by category

Justification category	Shoreline				Marine life				Natural order: Eval.	Endang. species: Assess. of value
	Assess. of value	Act eval.	Rule conting.	General.	Assess. of value	Act eval.	Rule conting.	General.		
Anthropocentric										
Personal interests	23	18	18	10	18	6	10	7	4	9
Welfare	43	37	38	48	34	31	27	45	10	29
Educative	1	1	0	0	1	0	2	2	0	6
Justice	2	0	2	4	2	2	2	2	0	0
Aesthetics	6	4	3	6	7	6	3	0	2	14
Biocentric										
Intrinsic value	2	0	2	2	8	12	10	9	35	13
Justice	1	4	2	2	11	25	13	12	10	3
Harm to nature	20	37	37	28	18	17	34	24	39	26

Source: Reprinted from Kahn (1997a), p. 1094.
Note: Percentages may not equal 100 because of rounding. Assess. = assessment; eval. = evaluation; conting. = contingency; general. = generalizability; Endang. = endangered.

tures. One researcher said: "[T]hey just got cooked by the sun. If you were to see them, you might think that these biota still were being killed by oil" (quoted by Matthews 1993, p. 7). Thus in public discourse and courts of law, important disagreements rest not only on establishing environmental harm but on whether such harm arises due to human activity or natural conditions. In the current study, we pursued this issue by asking children the following: "In nature, fish often eat other fish in order to live. Thus in nature many fish get killed. Is this different from fish dying in an oil spill? Why?" Results showed that most of the children (well over 90%) believed different issues where at stake for whether human activity (an oil spill) or natural predation caused the death of marine life. Here are examples of two common justifications:

Because when they're eaten by fish they are eaten by fish to keep them alive. And when they die in an oil spill they just die in the pollution and they can't really be used for anything. But when they are eaten [by people] *they are used to keep other people alive.*

[In an oil spill, the fish] *really don't have a chance to get away, to defend themselves. It just comes. But, see, if it's another fish, they can run or something. They can at least try to get away. You see, because if you waste it, then another fish can't eat it neither.*

Both justifications appeal to a system of dependent relationships, which in turn appears to establish the concept of a natural balance (e.g., "because if you waste it, then another fish can't eat it neither").

Telos of Nature

Aristotle (1962) begins *Nichomachean Ethics* by asserting that "the good, therefore, has been well defined as that at which all things aim" (p. 3). Aristotle then develops a teleological account of the good, wherein each kind of inanimate object (e.g., a ship) and animate being (e.g., a human) has an ideal way of functioning. Moreover, in humans, according to Aristotle, all the excellences of moral character can fit together into a harmonious self (Williams 1985). Something of this Aristotelian orientation emerged from the data. Consider two examples:

Yeah, because it looks better. . . . Well, I mean without any animals the world is, like, incomplete, it's like a paper that's not finished.

There's people, nature, and animals. That's what I think makes up the environment. And you're killing one-third of the environment that way. I don't think that's right. If you have a car, and you have everything but the motor, it's no good; it's kind of like the environment.

Both children offer a moral conception of the proper endpoint of nature, and that the good arises with nature reaching that end and being complete ("without any animals the world is, like, incomplete, it's like a paper that's not finished"; "If you have a car, and you have everything but the motor, it's no good; it's kind of like the environment").

The Development of Biocentric Reasoning

In the Houston child study, I had proposed that in the course of late childhood biocentric reasoning arises through the hierarchical integration of anthropocentric reasoning. Based on this proposition, I had expected in this current study (which included somewhat older children) that with increasing age the use of biocentric reasoning would increase while the use of anthropocentric reasoning would decrease. But the results showed a different pattern: as expected, biocentric reasoning increased with age, but so did anthropocentric reasoning.

Let me explain what I think is going on. Recall from chapter 3 one of the classic conservation tasks used by Piaget. A child is presented with two identical beakers filled with the same amount of water. In front of the child, the experimenter pours the water from one of the beakers into a third beaker that is taller but thinner. The empty beaker is then removed from the child's sight. The two beakers with water are placed next to each other. The experimenter now asks the child a pivotal question: Does the first beaker have more water, does the second beaker have more water, or do both beakers have the same amount of water? Simply stated, the conserving child (often only by five or six years of age) can understand that while one of the beakers of water is taller, it is also thinner, and can simultaneously set the two attributes in a compensatory relation. Now, in certain contexts, the same child neither needs nor employs such compensatory operations. Imagine, for example, if the child was asked to choose the beaker that could best help her reach an object outside her grasp. Likely, she will choose the taller one, because length particularly matters for this second task. We could ask her, "Why did you choose that beaker?" She might answer, "Because it is taller." Such a response would not be evidence that this child cannot set height and width into a compensatory relationship. The task did not require it.

Along a similar line of reasoning, it is possible that through development unelaborated welfare concerns give way to both human-oriented and nature-oriented considerations. We can initially think of this developmental pathway as a "Y." The bottom stem represents unelaborated welfare concerns; and the two parts above the split represent human-oriented and nature-oriented considerations. The mental organization of each group of considerations initially can be considered a structure. But in development I propose that the biocentric structure comes to encompass these two earlier structures (that we can now think of as partial structures) and their subsequent coordinations. Note, then, in accord with the data and the above Piagetian analogy, that this account does not preclude the seemingly isolated use of the partial structures.

Based on this account, it follows that one of the central places to look for the development of biocentric reasoning is in these coordinations between human-oriented and nature-oriented considerations. Here is an example. As described in table 8.1, aesthetic reasoning was defined as an appeal to the preservation of the environment for the viewing or experiencing pleasure of humans. Understood in this way, aesthetics is a human-oriented justification. But it also appeared from the data that some biocentric concepts involved valuing the natural environment in some experientially aesthetic way. For example, recall the above teleological justification: "Yeah, because it looks better. . . . Well, I mean without any animals the world is, like, incomplete, it's like a paper that's not finished." Here aesthetic reasoning ("it looks better") is linked to a biocentric conception of the proper endpoint of the natural world. Thus it is possible that an aesthetic sensibility helps foster the development of biocentric conceptions of virtue and value (compare Jarrett 1957, 1991; Nevers, Gebhard, and Billman-Mahecha 1997; Thompson 1995).

Coordinations between human-oriented and nature-oriented considerations can also be found in the four ways that children established biocentric justice reasoning: by establishing a direct relation between humans and nature, a compensatory relation, a conditional relation, and a hypothetical perspective-taking relation. As portrayed in table 8.1, an example of a direct relation is embodied in the following reasoning of a fifth-grade girl, Jill: "I think fish and animals have a right to live just like we do; and it's not fair to have killed them this way." Such reasoning is a coordination because she directly equates nature and humans.

Moreover, these four forms of reasoning may reflect a developmental progression. Imagine, for example, if we had asked Jill, "what if a person really loved the taste of fish, would it then be all right for that person to eat fish?" Let us say that Jill says, "No, fish have a right to live just like we do." Here she would be resisting the personal counterclaim (as discussed in chapter 4). We then move to moral counterclaims: "What if a person had a health problem that improved by eating fish?" Jill could say, "No, fish have a right to live just like we do." We ask, "What if a person was on a deserted island and this was her only way to live?" Again, Jill could say, "No, fish have a right to live just like we do." Indeed something like this very conversation occurred with Arnold, as discussed in the Houston child study. Recall that Arnold resisted countersuggestions that one could kill a mosquito with moral sanction. I suggested then that when such counterclaims gain purchase in an individual's psyche, the claims initiate the disequilibratory process that leads to development.

How, then, do we characterize more advanced biocentric reasoning? Consider what an eighth-grade girl, Mary, says: "You put yourself in the animal's position and you wouldn't like that, and so if you just kind of trade places and think about it and everyone would think it wasn't right." Here Mary engages in Kantian-like principled reasoning, wherein impartiality and generalizability are organizing features of her moral judgment. Such categorical reasoning—and this point is often misunderstood in critiques of Kant, Rawls, Kohlberg, and others—actually allows the individual to be "noncategorical" in the sense of not rigidly adhering to a single, concrete course of action. For example, when the nature writer Pam Houston (1996) walks in grizzly country, she willingly accepts that she is prey for a predator, just as she at other times—with moral sanction—is the predator to prey. In other words, Houston appears to do what Mary suggests: she puts herself in the animal's position, and accepts killing in a certain context. For such reasons, I think that a hypothetical equivalency (Mary's reasoning) is structurally more advanced than a direct equivalency (Jill's reasoning).

Structural Linkages between Justice, Welfare, and Virtue Reasoning

Over a decade ago, one of the more heated controversies in the moral developmental literature centered on Gilligan's (1982) claim that men and boys are oriented toward justice, and women and girls toward an ethic of care and responsibility. In chapter 4, I said that Gilligan's initial claim has not held up, and I suggested that part of the difficulty in the claim was that Gilligan had sought to separate what may be two interdependent constructs: justice and welfare. I had then sketched some of these possible linkages in the context of my own research on children's obligatory and discretionary moral judgments.

In the current study, similar linkages appeared in children's environmental moral reasoning. Indeed justice and welfare reasoning at times appeared so intertwined that we had difficulty in teasing them apart for purposes of coding. Consider one such example:

Like an animal, you can't just make one that's exactly like it was before— that's like a human being. They have brains, they're alive, they suffered just like we suffer, like an animal—we are an animal.

This child establishes a direct equivalency between humans and animals ("that's like a human being," "we are an animal"). On this basis, we considered coding the reasoning as biocentric justice, capturing the idea of reciprocal and fair relations. But the child also establishes a direct concern for the welfare of life ("they suffered") and for the intrinsic value of life ("you can't just make one that's exactly like it was before"). Thus, in our reading of this segment, the intrinsic value of life appears primary, while the value considerations themselves perhaps coalesce through a justice structure. Accordingly, we coded this reasoning as intrinsic value of nature. Granted, it is a difficult demarcation to make. But such difficulties in demarcating justice, welfare, and virtue reasoning would at times be expected given the proposition I have advanced in this chapter: namely, that biocentric reasoning arises through these very coordinations.

Children's Conceptions of Living in Harmony with Nature

Finally, we pursued a line of investigation initiated in the Houston parent study. We asked children what it means for them to live in harmony with nature, and to provide examples. Children's conceptions were coded with the categories reported in table 8.3. Results showed three developmental effects. First, in comparison with second- and eighth-graders, fifth-graders more often conceived of harmony in terms of positive acts. Second, compared with second-graders, eighth-graders more often conceived of harmony in terms of experience with nature. Third, in comparison with second-graders, fifth- and eighth-graders more often conceived of harmony in terms of "respect for nature" and "balance with nature" (which were grouped together for the statistical analysis).

In chapter 4, I reported on an earlier study I conducted on children's conceptions of trust (Kahn and Turiel 1988). One of the findings—which I said I would come back to in this chapter—was the developmental trend for children's conceptions of friendship. First-graders largely conceived of friendship in particularistic terms: particular events, personal predilections. For example, a friendship would be undermined "because I wanted him to play with me," "because he did something I don't like," or "because she lied to me." Third-graders began to conceptualize friendship from the perspective of ongoing relationships. Stability was based on quantitative factors, such as the number of times a violation was committed ("she would still be my friend as long as she didn't do it a lot of times") or the magnitude of the consequences ("like if he stole a clock from me, or something [more valuable than the stimulus presented], then I would be mad and he would drop a little"). Fifth-graders began to integrate quantitative factors with the issue of how a violation bears on the quality of the relationship, especially in terms of reliance, trust, and reciprocity ("all those years of being friends and stuff, why let it go away in one day").

In an intriguing sense, this developmental shift would seem to parallel the shift in children's conceptions of harmony just reported. In both, there is a movement from a focus on concrete reasoning about immediate events to more long-term, integrative, and abstract reasoning. Possibly both studies tap into the same concepts about relationship: in the trust study in terms of a human relationship and in this current study in terms of a relationship with nature. But I am hesitant to push this comparison too far. It could be said, for instance, that these findings follow in line with many other developmental studies that show similar shifts in chil-

Table 8.3
Children's conceptions for living in harmony with nature—summary of categories

Category	Basis of conception
Possessing	Having or possessing aspects of nature ("[harmony means] flowers in your yard, a big house")
Acting upon	Doing something to or for nature, including positive and negative acts ("[harmony] would mean you're doing great with not littering, not polluting the air")
Experiencing	Experiencing or interacting with nature ("[to live in harmony] means to have the experience of coming in contact with nature; go in the woods and have an experience, like go out there for a couple of days—go camping")
Educative	Conception based on learning from nature ("[to live in harmony] means to learn from the animals; if we start to look at animals as an example, then maybe we can change ourselves")
State of mind	Experiencing a particular state of mind or feeling ("[you know someone is living in harmony with nature] by their expression toward life, how some people wake up in the morning and say, 'It's a glorious day,' they love that day, and they are happy")
Relational	Relationship between humans and nature (it's just like it's your only child you can have, you love your child, it's just like loving animals like dogs and all of that")
Respect for nature	Respecting nature ("[harmony] means to respect lower life forms and respect animals when you see animals")
Balance with nature	Being in balance with nature, through a sense of either proportion or equality ("it's important to keep the scales balanced, taking from nature small amounts, but putting back what you take")

Source: Reprinted from Kahn (1997a), p. 1094.

dren's thinking across diverse content areas, such as children's conceptions of death (Childers and Wimmer 1971), kindness (Youniss 1980), authority (Damon 1977), groups (Niederman 1978), and person (Barenboim 1975). Accordingly, some theorists have suggested that there exists a global mental structure that cuts across different areas of knowledge (Case 1992). Such a characterization has its corollary in popular discourse. It is common, for example, to hear talk about how young children are concrete thinkers, moving only in adolescence and adulthood to the world of abstract thinking and abstract knowledge.

On a general level, such characterizations of development are correct, but so general that I worry that they are uninformative, if not misleading. Is it really true, for example, that young children do not engage in abstract thinking? Consider a one-year-old child who follows a ball as it rolls along the floor. The ball rolls under a couch and passes out of sight. The one-year-old then crawls around the couch to retrieve the ball. In Piagetian terms, this child appears to have the concept of object permanence. But think of what this means. The one-year-old believes an object exists even when he or she cannot actually see it. That is abstract reasoning. Of course, it is a different form of abstract reasoning than being able to understand the implications of a logical syllogism. But that is my point. What is interesting—in just the way Piaget understood—are the characterizations of the developing structure of knowledge. I raise this issue as a word of caution for future developmental research in this field. As this research continues, I suspect more cases will be found of some global patterns of reasoning. These findings will be of interest; but I do not think they should guide the theoretical underpinnings of a research program.

Conclusion

This study examined children's moral and ecological reasoning about the Prince William Sound oil spill. Results showed that most of the children understood that the oil spill negatively affected the local Alaskan shoreline (nonsentient nature), marine life (sentient nature), fishermen, recreationists, and the oil company responsible for the spill. It mattered to the children that harm occurred to the shoreline and marine life. Children also distinguished between death wrought to marine life by human activity (an oil spill) and by predation by other marine life. In consort with moral justifications, children's moral obligatory judgments were assessed using three criterion judgments: prescriptivity, rule contingency, and generalizability. Results showed that the majority of children—although less so for the second-graders—conceived of the harm caused to the shoreline and marine life as a violation of a moral obligation. We also asked children what it means for them to live in harmony with nature, and to provide examples. Results showed a developmental trend. Younger children more often conceived of living in harmony with nature in terms of a particular type of right action, while older children also incorporated conceptions of respect for nature and living in balance with nature.

In addition to providing specificity on children's moral environmental reasoning about an environmental disaster, these results help us gain a new perspective on an important developmental question: In ontogeny which comes first, moral considerations for humans (which then is applied to nature) or moral considerations for nature (which then is applied to humans)? Based on the results from this current study, I think the answer is neither. Rather, I think there is a dialectical relationship wherein children's moral relationships with other humans help establish their moral relationship with nature, and vice-versa. More technically, unelaborated welfare concerns appear to give way to both human-oriented and nature-oriented considerations (partial structures), which then undergo coordinations in development. In turn, one of the central places to look for the development of biocentric reasoning is in these coordinations. And that is just what we did when examining how an aesthetic sensibility may help foster the development of biocentric conceptions of virtue and value, and when examining the development of biocentric justice reasoning.

Overall, what is at stake is how to pursue developmental research that is sensitive to the ways in which children differentiate forms of knowledge while, by and large, living an integrated life. Differentiations and integrations: that is the story line here.

If my developmental account is even roughly correct, then it shows a fundamental limitation in young children's relationship with nature.

Namely, we could conclude that young children are not biocentric. But just as I argued earlier that we want to be cautious in accepting global characterizations of the mental life of children, so do we need to be cautious in interpreting such a conclusion. Young children are not biocentric as defined by the coordinations analyses I have sketched, but this does not preclude other forms of intimate affiliations that young children may have with the natural world.

9

The Brazilian Amazon Study

I start with a story from the Talmud:

Two men were fighting over a piece of land. Each claimed ownership and bolstered his claim with apparent proof. To resolve their differences they agreed to put the case before the rabbi. The rabbi listened but could not come to a decision because both seemed to be right. Finally he said, "Since I cannot decide to whom this land belongs, let us ask the land." He put his ear to the ground and, after a moment, straightened up. "Gentlemen, the land says it belongs to neither of you—but that you belong to it."

Compare this story to an often quoted sentence from a speech Chief Seattle purportedly wrote: "The earth does not belong to us, we belong to the earth." Very different traditions. One Hebraic, the other Native American. Yet both traditions speak to what I have called a biocentric relationship with nature, where nature is understood as not merely serving human ends, although it can, but having moral standing in its own right and forming part of a larger community in which humans live.

In both the Houston child study and Prince William Sound study, I showed that children can have biocentric conceptions. Yet the children in both studies provided such conceptions infrequently. For example, in the Houston child study biocentric reasoning comprised only 4% of children's total justifications. In turn, I had interpreted this finding from a developmental perspective as suggesting that biocentric reasoning begins to emerge, if at all, in late childhood.

However, an alternative interpretation is possible, and it hearkens back to Nelson's (1993) perspective on biophilia, quoted in chapter 1.

[M]uch of the human lifeway over the past several million years lies beyond the grasp of urbanized Western peoples. And if we hope to understand what is fundamental to that lifeway, we must look to traditions far different from our own. . . . Probably no society has been so deeply alienated as ours from the community of nature, has viewed the natural world from a greater distance of mind, has lapsed to a murkier comprehension of its connections with the sustaining environment. Because of this, we are greatly disadvantaged in our efforts to understand the basic human affinity for nonhuman life. Here again, I believe it's essential that we learn from traditional societies, especially those in which most people experience daily and intimate contact with land. . . . " (pp. 202–203)

Thus it could be said that in our previous studies we found so little biocentric reasoning because urbanized Western people, children included, have largely lost their biocentric connections with nature. In turn, following Nelson, one would expect to find a higher percentage of biocentric reasoning with more indigenous people who, as Nelson says, experience daily and intimate contact with the land.

In the current study, colleagues and I pursued this line of inquiry (Howe, Kahn, and Friedman 1996; Kahn 1997c). We modified the methods from the Houston child study and interviewed, in Portuguese, 44 fifth-grade Brazilian children in urban and rural parts of the Amazon jungle. Thirty children (16 females and 14 males) came from Manaus (mean age, 13–8), the capital of the State of Amazonas, and 14 children (7 females and 7 males) from Novo Ayrão (mean age, 13–7), a small remote village.

Since both geographical locations may be unfamiliar to the reader, I start with a few words about both. Then I focus on the quantitative results that begin to portray the environmental views and values of the children in both locations. I discuss children's environmental justifications and portray children's feelings of both pessimism and optimism in their ability to effect meaningful environmental change. Next, I compare many of these results to those from the Houston child study. Finally, taking all of these results together, we will be positioned in the conclusion to evaluate the above proposition: that through intimate contact with the land children develop a deeper affinity for it.

The Two Locations

With nearly one million inhabitants, Manaus is the largest Brazilian city within the vast Amazon rain forest. The city is located thirteen miles above the junction of the Rio Negro and the Amazon River, and it is at this junction that the Amazon River is said to begin. Manaus services a growing ecotourist trade from North America and Europe. The city is also considered the center of the region's electronics industry, and it enjoys tax-free imports due to the government's efforts to spur international development in the region. Yet, even given this economic development, for many Brazilians Manaus remains an emblem of the economic disillusionment that plagues Brazilian society (Nyop 1983; Potter 1989). At the turn of the century, approximately 90% of the world's rubber supply passed through the city, resulting in rapid economic and cultural growth. When the market crashed in 1925, this growth came to an abrupt and debilitating halt (Beresky 1991). Currently, within Manaus a great deal of poverty exists, as do poor medical care and a lack of educational and job opportunities. In some sections of the city, refuse and litter are readily apparent, and sickness manifests (e.g., cholera, malaria, and yellow fever). The urban children who were interviewed attended a school in São Raimundo, a neighborhood of only modest economic means in comparison to the city as a whole. Some of these children lived near creeks that some people used as their primary means for garbage and sewage removal.

In contrast, Novo Ayrão is a village of approximately 4000 inhabitants. The village could only be reached by means of an eight-hour boat ride up the Rio Negro from Manaus. The villagers' primary economic activities included fishing and the extraction of forest products, most notably lumber. The landscape is largely pristine with only small areas cleared for housing, commerce, and dirt roads. There is little visible litter or garbage, and according to some inhabitants neither crime nor drugs are present in the community. The children who were interviewed attended one of the village's two schools.

Children's Environmental Profile

Paralleling the Houston child study, we asked questions to ascertain in what ways children were aware of environmental problems, believed that certain human activities caused environmental problems, acted to solve some of these problems, and cared for aspects of nature. The quantitative results are reported in table 9.1. As shown in this table, virtually all the children in both locations said animals and plants played an important part in their lives. The majority of children were aware of environmental problems that affected themselves or their community, and spoke of concerns that focused on plants and forests, such as the large-scale burning of the Amazon jungle (53% and 56% in Manaus and Novo Ayrão, respectively), air pollution (24% and 33%, respectively), harm to animals (12% and 11%, respectively), and garbage or litter (12% and 0%, respectively). The majority of children discussed environmental issues with their families. The children in this latter category said they talked about plants and forests (46% and 47%, respectively), animals (29% and 27%, respectively), air pollution (14% and 7%, respectively), water pollution (0% and 7%, respectively), and garbage and litter (0% and 7%, respectively). Statistically, children in Novo Ayrão more often said that they acted to help solve environmental problems than did children in Manaus. Children's reported environmental actions included planting trees or in some way caring for plants and trees (50% and 85%, respectively), caring for animals (21% and 15%), and influencing other people to be environmentally responsible (21% and 0%, respectively).

In another series of questions, we asked children to imagine that their entire community threw garbage in the Rio Negro. As shown in table 9.1, in this situation the majority of children believed harmful effects would result for birds, insects, the view, and the people living along the river. In addition, of those children who believed that such harm would occur, the majority said it would matter to them if such harm occurred.

To provide an overall assessment of these children's environmental profile, and to test in one place for effects of location (Manaus vs. Novo Ayrão), ten of the above questions were summed as a single score, reflecting the degree of each child's pro-environmental views and values. The questions included those that pertained to whether children said animals and plants were an important part of their lives, were aware of environmental problems generally, discussed environmental issues with their family, initiated a conversation about nature, acted to help the environment, and cared about aspects of nature (birds, insects, the view, and people along the river). For each question, an affirmative response received a score of one, a negative response a score of zero, and then the scores were summed across the ten questions. Results showed that out

Table 9.1
Percentage of children's environmental values, knowledge, and practices

Environmental criterion	Manaus ($n = 30$)	Novo Ayrão ($n = 14$)	Houston ($n = 24$)
Animals an important part of your life.	100	100	91
Plants an important part of your life.[a]	97	100	79
Aware of environmental problems in general.	69	57	—
Aware of environmental problems affecting self and community.	81	86	80
Discuss environmental issues with family.	62	64	71
Initiate discussions on environmental issues.	31	43	—
Act to help solve environmental problems.[b]	41	79	—
Thinks that throwing garbage in a river harms birds.	97	86	96
Cares that birds would be harmed.	96	100	95
Thinks that throwing garbage in a river harms insects.	57	64	68
Cares that insects would be harmed.[a]	61	58	89
Thinks that throwing garbage in a river harms the view.	97	100	91
Cares that the view would be harmed.	93	92	95
Thinks that throwing garbage in a river harms people along the river.	93	100	95
Cares that people would be harmed.	89	85	81

Source: Howe, Kahn, and Friedman (1996), p. 983.
Note: Children were first asked if they thought harm occurred (to birds, insects, the view, or people). Only those children who thought harm did occur were then asked if they cared about the harm. The dash indicates that a comparable question was not asked of the Houston children.
a. There was a statistical difference between the population in Brazil (Manaus and Novo Ayrão) and the United States (Houston), $p < .05$.
b. There was a statistical difference between the populations in Manaus and Novo Ayrão ($p < .05$).

of a possible score of 10 (the most pro-environmental score), the Manaus children as a group scored 7.03 and the Novo Ayrão children as a group scored 7.50. Statistically, there was no significant difference between the two groups.

Children's Understandings of and Values toward the Amazon Rain Forest

By means of large-scale logging and burning, humans are destroying the Amazon rain forests. What do the children who live within this region think of this situation?

We found the following. All the children (100%) in both Manaus and Novo Ayrão believed that humans need the forest, and virtually all the children (93% and 100%, respectively) could name at least one thing the forest provided. Children in this latter category said the forest provided food (33% and 32%, respectively), clean air or oxygen (17% and 19%, respectively), lumber (7% and 16%, respectively), medicine (17% and 7%, respectively), animals (10% and 10%, respectively), shade (2% and 10%, respectively), and beauty (5% and 0%, respectively). In turn, the majority of children in both Manaus and Novo Ayrão believed that people are currently cutting down the rain forest (83% and 100%, respectively), that such actions are wrong (83% and 79%, respectively), that the government should stop the people who are cutting down the rain forest (96% and 92%, respectively), that they themselves should take some action to help stop the cutting of the rain forest (88% and 85%, respectively), that there is a way to use the forest without destroying it (72% and 86%, respectively), and that the rain forest will exist forever (64% and 64%, respectively).

To provide an overall assessment of these children's conservation views and values, and to test for effects of location (Manaus vs. Novo Ayrão), all of the five above questions that pertained to conservation were summed as a single score. One question focused on whether humans need the forest, two questions on children's awareness and judgments of current logging practices, and two questions on children's judgments about personal and governmental interventions. This process was similar to that employed in computing the environmental profile. For each question an affirmative response received a score of one, a negative response a score of zero, and then the scores were summed across the five questions. Results showed that out of a possible score of 5 (the most pro-conservation score), the Manaus children as a group scored 4.0 and the Novo Ayrão children as a group scored 4.20. Statistically, there was no significant difference between the two groups.

Many of the children's fathers in Novo Ayrão gained their livelihood through logging; yet still, as shown above, the majority of the Novo Ayrão children believed such activity was wrong. Here is an example of a child's dialogue with the interviewer:

Do you think that people are cutting down the jungle? *Yes, very much so. Mainly my father.* Do you think that is right or wrong? *Yes. Because it destroys it.* And why is that wrong? *Because we need the jungle, it purifies the air.* Do you think the government should stop these people? *I can't say anything because my father works cutting down the trees. In my opinion it should.*

On the one hand, this child appears resistant to criticize her father ("I can't say anything because my father works cutting down the trees). On the other hand, she judges her father's actions as wrong and says the government should stop the cutting ("In my opinion it should"). Other times children legitimated the economic need for logging while maintaining their evaluative judgment against it. For example, one child said: "[The government shouldn't stop people from cutting trees] because some people make their living out of selling the trees they cut. . . . I think it is wrong, but they need to do it, it is their job. That is how they bring food home. . . . "

The Case of the Polluted Waterway

Paralleling the Houston child study, we investigated children's moral views toward throwing garbage in their local waterway—in this case the Rio Negro. Keep in mind that the common practice for some residents within Manaus was to discharge their garbage by this means. Evaluative results showed that virtually all the children interviewed in both Manaus (97%) and Novo Ayrão (93%) judged the individual act of throwing garbage in the Rio Negro as not all right. Children maintained their judgments about not throwing garbage in the river even in conditions where local conventions legitimated the practice for their entire community (97% and 93%, respectively). Children also generalized their prescriptive judgments to a different geographical location that adhered to that common practice (93% and 86%, respectively).

We then examined whether children provided negative evaluations on each of these three measures: prescriptivity, noncontingency on conventional practices, and generalizability. Taking this result as an initial measure of moral obligation, we found that 93% of the children in Manaus believed throwing garbage in the Rio Negro violated a moral obligation, compared to 86% of the children in Novo Ayrão. Statistically, there was no significant difference between the two groups. In the next section, I will couple this analysis with the justification data.

Children's Justifications

We systematically pursued children's justifications for six of their evaluations. Two evaluations involved whether animals and plants played an important part in their life, three involved "The Case of the Polluted Waterway," and one involved whether it was all right or not all right to cut down the Amazon forests. Children's justifications were coded with the categories reported in table 9.2. The resulting justification percentages for each of the six questions, separated by location (including the Houston location, which will be discussed shortly), are reported in table 9.3. Averaging across all six questions, results showed that the majority of Brazilian children's justifications were anthropocentric (77% and 79% for Manaus and Novo Ayrão respectively), followed by unelaborated harm to nature (19% and 13%, respectively), and then biocentric (4% and 8%, respectively). Based on these three overarching categories, results showed no statistical difference in justification use between urban and rural groups. However, a subsequent analysis split the general anthropocentric category into two subcategories: welfare and other. There was a significant statistical effect for group, subsequently revealing that in comparison to the urban children (57%), the rural children (67%) used a greater proportion of welfare justifications.

I said above that the large majority of children believed that throwing garbage in the Rio Negro violated a moral obligation, and that (as discussed in chapter 4) I would couple that analysis with the justification data. I can do that now. For children who evaluated as not all right polluting the river in all three conditions, we conducted an analysis that examined the percentage of children who provided moral justifications for their negative evaluations. Results showed that all the children (100%) provided a moral justification (i.e., a justification coded as welfare, intrinsic value of nature, rights, relational, and/or unelaborated harm to nature) for at least one of their three evaluations (90% for two of the three, 58% for three of the three).

At this point, I should emphasize what I noted in the Houston parent study—that over the years I have been puzzled by the relational justification category. Let me recount the basic sequence of events. Early on, colleagues and I established the coding category labeled personal interests, which is a fairly standard category in the social cognitive literature (see chapter 3). We defined this category as an appeal to personal interests and projects of self and others, including those that involve recreation or provide fun, enjoyment, or satisfaction. For example, in the Prince William Sound study, a child said that the oil spill was wrong because "people go to beaches for fun, not just to sit there and look at oil everywhere." But then we started noticing relational aspects of this reasoning, particularly in the context with animals. For example, one child said, "Pokey [my dog] is my friend because he's very close, like my relatives, and he's very fun to be with." Here we thought that this child largely grounded her relationship with her pet in terms of personal interests ("he's very fun to be with"). But we still remained puzzled by the relational considerations in terms of friendship and psychological rapport. We then noticed other instances where such relational considerations became more central. Another child said, "Animals are important to me because when a person in my family, like, died, they could come and cheer me up." Granted, on the one hand, this reasoning still refers, if not exactly to personal interests, at least to an anthropocentric form of reasoning, where the animal serves the needs of the child. But, on the other hand, this reasoning also refers to having a relationship with

Table 9.2
Summary of environmental justification categories

Category	Summary
Anthropocentric	An appeal to how effects to the environment affect human beings. In other words, the environment is given consideration, but this consideration occurs only because harm to the environment causes harm to people.
Personal interests	An appeal to personal interests and projects of self and others, including those that involve recreation or provide fun, enjoyment, or satisfaction (e.g., "animals are important; for instance, in the zoo there are a lot of people who like to see the animals, like myself"; "I think the jungle offers fun; for example, go camping during the weekend").
Aesthetic	An appeal to preservation of the environment for the viewing or experiencing pleasure of humans (e.g., "plants are important because they give up a good smell, they are beautiful, very pretty"; "because sometimes we are mesmerized with the beauty of the jungle"; "rivers that are polluted, full of trash, are very ugly").
Welfare	An appeal to the physical, material, and psychological welfare of human beings (e.g., "we should preserve the plants and not destroy them because it brings us oxygen and we can survive through it"; "because it causes pollution that is dangerous for us, because now we have cholera, a very dangerous disease, and there are others attacking us like malaria").
Punishment avoidance[a] Unelaborated	An appeal to punishment or its avoidance (e.g., "because the police might catch her").
Biocentric	An appeal to a larger ecological community of which humans may be a part.
Intrinsic value	An appeal that nature has value, and the validity of that value is not derived solely from human interests, including is-to-ought appeals (e.g., "because the river was not made to have trash thrown in it, because the river belongs to Nature"; "because the jungle, God made it to live and not to be cut").
Rights	An appeal that nature has rights or deserves respect, including appeals wherein humans and nature are viewed as essentially similar (e.g., "because the animals think like us"; "because birds have a life as we do, they have a mother, they are like us"; "plants are born, reproduce, and die as we human beings do").

Table 9.2 (continued)

Relational	An appeal to a relationship between humans and nature, including those based on psychological rapport (e.g., because the animals are our friends") and stewardship (e.g., "plants are important to me because we should take care of them, but a lot of people don't do it, they cut them down, so we have to preserve nature"; "because the jungle can't defend itself, somebody has to defend her").
Unelaborated harm to nature	An appeal to the welfare of nature (e.g., "because the birds need the water of the rivers to drink, and if it gets polluted it kills many birds and animals"; "because it is going to kill the fish, the river is going to be polluted"; "because they are destroying the Amazon jungle"). No reference is made to whether that concern derives from an anthropocentric or biocentric orientation.

Source: Reprinted Howe, Kahn, and Friedman (1996), p. 982.
Notes: a. Although virtually none of the Brazilian children used this justification category of punishment avoidance, we included it in the coding manual in order to be particularly sensitive to any moral or environmental orientations that were based on punishment. The example comes from the Houston child study (Kahn and Friedman 1995).

an animal, such that an animal could provide psychological comfort. At that point we generated a new category called "relational" which in this study we subsumed under the biocentric category. But I still am not sure. Indeed, in the Portugal study (chapter 10) we reverse our decision once again and subsume relational reasoning under the anthropocentric category.

Children's Feelings of Pessimism and Optimism in Effecting Environmental Change

In many countries, people can feel pessimistic about ever effecting meaningful environmental change. In the former Soviet Union, for example, many people feel powerless as they lose their livelihoods and their health due to the region's massive environmental problems (Peterson 1993; Russians Struggle 1994). In the United States, Thomashow (1995) reflects on "20 years of teaching and innumerable discussions about the fate of

Table 9.3
Percentages of environmental justification by categories

Justification category	Important part in your life						Case of the polluted waterway									Rain forest		
	Animals			Plants			Individual			Local community[a]			Distant community[a]			Deforestation		
	M	N	H	M	N	H	M	N	H	M	N	H	M	N	H	M	N	H
Anthropocentric																		
Personal interest	21	7	35	3	5	8	8	6	8	0	0	6	4	7	3	0	0	—
Aesthetic	17	0	4	27	29	33	0	0	31	7	0	19	7	0	19	26	6	—
Welfare	38	64	26	70	52	42	64	75	18	64	81	25	59	79	19	45	56	—
Punishment avoidance	0	0	0	0	0	0	0	0	3	0	0	3	0	0	0	0	0	—
Unelaborated	0	0	0	0	0	0	0	0	0	4	6	3	0	0	11	0	0	—
Biocentric																		
Intrinsic value	0	7	0	0	0	0	3	0	3	0	0	3	0	0	0	3	0	—
Rights	14	14	9	0	10	0	0	0	3	0	0	3	0	0	0	0	6	—
Relational	3	7	9	0	5	0	0	0	0	0	0	0	0	0	6	0	0	—
Unelaborated harm to nature	7	0	17	0	0	17	25	19	40	25	13	39	30	14	42	26	33	—

Source: Reprinted Howe, Kahn, and Friedman (1996), p. 984.

Note: M = Manaus (*n* = 30); N = Novo Ayrão (*n* = 14); and H = Houston (*n* = 24). Percentages may not equal 100 because of rounding. A dash indicates that this issue was not investigated in the Houston study.

a. Given the common practice to pollute.

the earth" through which his students sometimes become "increasingly depressed, feeling victimized, paralyzed, either blaming the externalized other or swimming in their own guilt" (p. 152). In Germany, Nevers, Gebhard, and Billman-Mahecha (1997) report that in several of their group interviews with children about the natural environment "a kind of fatalism has been observed, in which the children or adolescents depict themselves as passive . . . with little influence" (p. 183). Along similar lines, in our qualitative data from Brazil, children's feelings of futility sometimes emerged. For example, one child said: "If I had someone who would help me. . . . I would like to do something [about the problems] but I can't. Very few people pay attention to nature." Another child said she wished it were possible to stop the people from cutting down the jungle, but "I think it is impossible to stop them."

Informally, colleagues from Brazil have told me that they thought the cynicism about government runs deep in the Brazilian psyche, deeper than in many other countries. That proposition may be true (compare Reiss 1990), but we also noticed in our data certain forms of children's optimism. For example, some children thought all that was required for change was the effort: "[Whether the jungle continues to exist] depends on the people that live there and want to protect it. If they make the effort to protect it, it will exist forever." Other children thought that the jungle could be protected by looking for alternative ways of living: "We need to look for a way of life that doesn't use wood . . . work in something else that won't destroy nature." Still other children looked toward dialogue as a means of reeducation: "For instance, we see someone cutting down a tree, we should talk to him so he will stop doing it. If we do that, we already have done our share. If someone else is burning it, we do the same." Also, as noted above, the majority of children (64% in both locations) believed the rain forest would exist forever.

As in most if not all societies, it appears that the children can bring fresh voices in critiquing and finding solutions to the harm wrought by previous generations. Perhaps these children's voices will stay fresh. Or perhaps their voices will die out as the children move further toward adulthood and its corresponding demands. These issues warrant further research. Still, at a minimum, our informal qualitative data—taken with our quantitative results reported above—show that educators have a tremendous amount to work with in terms of these children's environmental sensitivities and commitments.

Comparison to the Results from the Houston Child Study

We collected the Brazilian data such that a good portion of it could be compared directly to the data from the fifth-grade black children in the Houston child study. Table 9.3 provides a quantitative comparison of the justification data. Statistically, we tested four categories for a group effect: anthropocentric welfare, anthropocentric other, biocentric, and harm to nature. Results showed that the children in the Brazilian populations (63%) used a greater percentage of anthropocentric welfare justifications than the children in the Houston population (23%). This effect may have arisen because rural and urban children in the Amazon region more directly depend upon nature for their physical survival than urban children in the United States. Indeed this explanation is supported by two group differences found in the evaluation data (table 9.1). First, more children in the Brazilian populations than the Houston population said plants played an important part in their lives. Second, fewer children in the Brazilian populations than the Houston population said they would care if insects (some of which carry deadly diseases in the Amazon region) were harmed by water pollution.

Otherwise, there were no further statistical differences across each of the twelve questions that pertained to children's environmental values, knowledge, and practices (table 9.1). Nor was there a statistical difference across children's environmental profile: the Houston children scored 7.8, compared to 7.0 and 7.5 for the Manaus and Novo Ayrão children respectively. In terms of the "Case of the Polluted Waterway," again there were no statistical differences. In brief, based on the three measures that pertain to moral obligation, 100% of the Houston children judged throwing garbage in a waterway as not all right in the three conditions, compared to 93% and 86% of the children in Manaus and Novo Ayrão, respectively.

When I first saw these results, I asked myself, "Why did we not find more differences?" My first response was that perhaps we had encountered a problem with a ceiling effect. That is, perhaps we had asked questions that would elicit a high percentage of affirmative responses across most any population, and thus we were unable to uncover important differences. But this concern is checked in two ways. First—and this is absolutely crucial—we had no ceiling effect based on our assessment of biocentric reasoning, and we still found no group differences. Second, even given a ceiling effect on some of the evaluation data, I think that the meaningfulness of the data stands. As an analogy, imagine a fifth-grade math teacher who teaches her students how to multiply two double-digit numbers. She then gives a test and finds that all her students solve at least 90% of the problems correctly. She, too, could be said to have encountered a ceiling effect; but the success of her students stands. It is the difference between norm-referenced and criterion-referenced evaluation. Similarly, I believe our evaluative questions provided meaningful criteria by which to conduct a comparison between these two populations.

My second response to the question of why so few differences was to ask whether I trust the Brazilian data. I have two hesitations. One is that the interviewer was only marginally fluent in Portuguese, and that made the probing difficult while interviewing. Still, the children were interviewed in their own language (Portuguese), which allowed them to speak their mind without a language barrier. Another hesitation is that while we had 30 participants in Manaus, which is fine, we only had 14 participants in Novo Ayrão. The small sample size in Novo Ayrão was not intended and arose from the various complications that can beset research in such cross-cultural settings. Still, the sample size was large enough to employ statistical techniques; and both Brazilian groups taken together (44 participants) made for a comparatively large sample to compare to the participants in the Houston child study.

Moreover, it is important to recognize that the coding system developed from the Houston child study was comprehensive enough to account for the Brazilian data. Indeed children's reasoning across both cultures sometimes almost echoed one another. For illustrative purposes consider the following four pairs of matched examples:

[It is not all right to throw garbage in the river] *because it causes pollution that is dangerous for us. Because now we have cholera, a very dangerous disease, and there are others attacking us like the malaria.* (Brazilian child)

Because some people that don't have homes, they go and drink out of the rivers and stuff, and they could die because they get all of that dirt and stuff inside of their bodies. (Houston child)

Both of the above children reason that it is wrong to throw garbage in the local waterway because people might drink from polluted water, and get sick ("now we have cholera, a very dangerous disease"; "they could die").

Because the river was not made to have trash thrown in it, because the river belongs to nature. (Brazilian child)

Because water is what nature made; nature didn't make water to be purple and stuff like that, just one color. When you're dealing with what nature made, you need not destroy it. (Houston child)

Both of the above children base their environmental judgments on the view that nature has its own purposes ("the river was not made to have trash thrown in it"; "nature didn't make water to be purple and stuff").

Because animals have to have their chance. They also must have to live. We should not mistreat them, because if it happens to us, we don't like it. (Brazilian child)

Some people don't like to be dirty. And when they throw trash on the animals, they probably don't like it. So why should the water be dirty and they don't want to be dirty? (Houston child)

Both of the above children judge as wrong the mistreatment of animals based on considering whether humans would similarly like to be treated in that way ("because if it happens to us, we don't like it"; "Some people don't like to be dirty . . . [so the animals] probably don't like it").

Even if the animals are not human beings, for them they are the same as we are, they think like we do. (Brazilian child)

Fish don't have the same things we have. But they do the same things. They don't have noses, but they have scales to breathe, and they have mouths like we have mouths. And they have eyes like we have eyes. (Houston child)

Both of the above children recognize that although animals are not identical to human beings ("animals are not human beings"; "fish don't have the same things we have"), both animals and people have significant functional equivalences (animals "think like we do"; fish "don't have noses, but they have scales to breath").

Conclusion

The majority of Brazilian children we interviewed demonstrated environmental sensitivities and commitments based on a wide range of measures. The children were aware of various environmental problems (such as air and water pollution, and the large-scale logging of the Amazon jungle). They discussed environmental issues with their families. They believed that throwing garbage in the Rio Negro hurt various parts of the environment (birds, insects, the view, and people who lived alongside the river), and they cared that such harm occurred. Based on the criteria of prescriptivity, noncontingency on conventional practices, and generalizability (and coupled with moral justifications), the children believed that throwing garbage in the Rio Negro violated a moral obligation. The children demonstrated understandings of and sympathies toward the Amazon rain forest, believing, for example, that the government should stop people who are cutting down the forests. The results also showed how children can hold to conflicting environmental judgments: in some cases wanting to endorse their parent's line of work (logging) but judging that work as wrong. Indeed it is such complexity of social thought, and its partial independence from the social milieu, that allows children, perhaps the world over, to critique and change the existing social order.

Through the interviews we obtained some instances of biocentric reasoning. Consider, for example, one girl's explanation for why the government should stop people from logging the jungle.

It is like me having a leg or an arm cut. . . . Nature is like a person, no, thousands of persons because it isn't just one thing. . . . [A] person is like a tree. If the tree bears fruits, it is the same with people. Taking care of a tree is the same. If you cut a branch off a tree it is like cutting a finger or the foot. To cut a tree down is like doing it to yourself. It is the same to our heart, it is not good. The jungle is like the heart of a person.

It is beautiful language: "The jungle is like the heart of a person." In turn, recall Nelson's (1993) perspective that "to understand the basic human affinity for nonhuman life" we must "learn from traditional societies, especially those in which most people experience daily and intimate contact with land" (p. 203). Following Nelson, one could well propose that since the children in Novo Ayrão lived in more intimate contact with the land than their Manaus and especially Houston cohorts, comparatively more of a biocentric orientation would have emerged in the Novo Ayrão children.

The results did not support this proposition. Between the rural children in Novo Ayrão and the urban children in Manaus there was no difference in the frequency of biocentric reasoning, and few differences overall. Even more startling, there were few differences between the Brazilian children (taking both groups together) and the black children in the Houston child study. Specifically, there were only two statistical differences across twenty-six separate questions (which formed a large body of both studies), no differences across the summed scored analysis that comprised the environmental profile, and no differences in the frequency of biocentric reasoning. In addition, the structure of children's reasoning often seemed identical.

If we cannot easily discount the results—and I argued earlier that we cannot—then why did we not find more differences, especially on the assessment of biocentric reasoning? One possible answer, provided informally by Roger Hart at the Graduate School of the City University of New York, is that while the village was accessible only by boat, it was still heavily influenced by the missionary culture; moreover, by interviewing the children in Portuguese (instead of an indigenous language), it could be said that our interview was weighted toward eliciting responses imbued with the missionary culture. Hart contends that had an indigenous population of Amazonian children been interviewed in their indigenous language a much higher percentage of biocentric reasoning would have been uncovered. Hart may be correct. But it still remains that the children in Novo Ayrão lived close to the land, with daily and intimate contact, and that by itself was not enough apparently to engender biocentric reasoning. Thus there may be something to Diamond's (1993) contention (see chap. 2) that biocentric reasoning does not emerge in every culture that lives close to the land. Still another possible answer follows from my tentative account of the development of biocentric reasoning offered in the previous chapter: that biocentric reasoning emerges, if at all, more fully in older adolescents and adults. If correct, then more biocentric reasoning might have been found with an older population in the village where we had conducted our research.

This single study only begins to broach the larger agenda of how to understand cross-culturally the human relationship with nature. I would but add here that toward addressing this question future researchers particularly need to be clear in defining their key terms, such as "affinity for nature," "affiliation with nature," or, in my case, "biocentrism." For without clarity problems arise in comparing across studies, as different people can mean different things by their terms, and indeed the same person can mean different things, as well. We encountered a version of this problem in chapters 1 and 2. There we started with a seemingly simple proposition, initiated by Wilson (1984): that humans have an innate propensity to affiliate with nature. But soon this proposition got complicated not only because of the claim that the affiliation has a genetic origin, but because we had difficulty knowing exactly what is meant by an affiliation. Does it mean that we find nature enjoyable? aesthetically pleasing? conducive to relaxation? necessary for our survival? or intrinsically valuable? Does it mean that we dislike certain aspects of nature? avoid it? and seek to dominate it? And if one says having an affiliation with nature includes all of the above and more, then does one really have left a meaningful basis by which to assess the proposition that native people have a greater affiliation with nature than urbanized Western people?

In responding to this problem, my approach has been to distinguish anthropocentric and biocentric environmental orientations, and then within each orientation to characterize various forms of reasoning in as much detail as possible, and their potential developmental pathways. I believe this approach can lead to substantive answers on how to understand both cultural differences and universal aspects in the human relationship with nature. In the next chapter, I extend this line of inquiry in an empirical study conducted in Portugal. Then, in chapter 11, I integrate this line of inquiry into a discussion on epistemology, culture, and the universal.

10

The Portugal Study

And when the woman saw that the tree was good for food, and that it was pleasant to the eyes, and a tree to be desired to make one wise, she took of the fruit thereof, and did eat, and gave also unto her husband with her; and he did eat. And the eyes of them both were opened, and they knew that they were naked; and they sewed fig leaves together, and made themselves aprons. And they heard the voice of the Lord God walking in the garden in the cool of the day: and Adam and his wife hid themselves from the presence of the Lord God amongst the trees of the garden. And the Lord God called unto Adam, and said unto him, Where art thou? And he said, I heard thy voice in the garden, and I was afraid, because I was naked; and I hid myself. And he said, Who told thee that thou wast naked?

Are humans natural? If not, is it because we have certain types of knowledge? Self-reflective capacities? Moral sensibilities? Such questions have a long history, as reflected in the above passage from Genesis (chap. 3, verses 6–11). Such questions also comprise one of the new lines of investigation in this study—the Portugal study (Kahn and Lourenço 1999).

In this study, we interviewed (in Portuguese) 120 students in Lisbon, Portugal, 30 students in each of four grade levels: fifth (mean age approximately 10–5; 30 females and 30 males), eighth (mean age approximately 13–6; 30 females and 30 males), eleventh (mean age approximately 16–7; 30 females and 30 males), and college (mean age approximately 19–4; 31 females and 29 males). In order to continue to make direct cross-cultural comparisons, we purposefully replicated from the Houston child and Brazilian Amazon studies a portion of our forty-two systematic interview questions. For example, we asked general questions concerning participants' environmental views and values and understandings of environmental problems. We also set up a scenario— "The Case of the Polluted Waterway"—wherein we examined, as in the other studies, moral obligatory judgments about water pollution. But we moved forward, as well, by extending the environmental content to include not just water (in terms of polluting a river) but three other categories of nature: air, fire, and earth. For air we set up a scenario— "The Case of the Driven Automobile"—wherein we investigated moral judgements about air pollution, and how participants coordinated personal interests with competing environmental claims. For fire (forest fires) we investigated participants' conceptions of the natural. With earth (logging) and fire (forest fires) we investigated how participants would solve environmental problems if they were empowered politically. Finally, overarching all these inquiries, we extended our age range into late adolescence and young adulthood so as to bring this population into our analyses of the human relationship with nature.

Appendix A includes the complete set of structured questions that we asked of every participant in this study. Appendix B includes our coding manual for the justification data. I have not presented this level of detail for any of the other studies so as not to burden the reader. But at least for one study such detail is warranted, for these appendices allow the reader (a) to scrutinize more deeply our methods and qualitative analyses, (b) to see what the transition looks like from developing a coding manual to providing a summary justification table (table 10.1), and (c) to utilize the coding system in related research.

An Initial Environmental Profile

As in our previous studies, we examined participants' views toward animals, plants, and parks, and awareness of environmental problems. Based on such measures, an initial environmental profile emerged. Specifically, participants said domestic animals (96%), wild animals (96%), plants (97%), and parks (100%) were important. Participants said they discussed environmental issues with family or friends (79%) and acted to solve environmental problems (90%). In addition, we assessed participants' judgments concerning consequences of throwing garbage in their local river—the Rio Tejo. Results showed that participants believed throwing garbage in the Rio Tejo would have harmful effects on fish (100%), birds (91%), water (100%), the view (98%), and people who live close to the river (100%). Of participants who believed harmful effects occurred, further results showed that it mattered to the participants if such harm occurred to fish (96%), birds, (95%), water (99%), the view (97%), and people who live close to the river (96%).

Companionship with Pets

Pets can provide affection, act as a confidant, and offer social contact. Moreover, as demonstrated by hundreds of clinical studies (reviewed in chapter 1), such interactions with pets can promote the physical and psychological health of people who are healthy or sick, young or old. With these ideas in mind, we examined the reasoning of the participants (96%, noted above) who said that pets were important. We found that 59% of their justifications appealed to companionship. Here are two examples:

[Pets are important] *because many of them become companions to those who own them, they are their friends, it is a help that is there to take care of them, it is company.*

[Pets are important] *because they are company, they are companions.*

Interestingly, only 1% of the justifications focused on how pets benefit the physical welfare of self or others, and only 4% of the justifications focused on how pets benefit the psychological welfare of self or others. Thus it would appear that while participants expressed a keen awareness of the relational aspect afforded by pets, they were less aware of (or at least less verbal in expressing) the resulting clinical benefits.

The Importance of Parks in Fostering Psychological Welfare

The word "park" translates best into Portuguese as "jardim," which then translates best back into English as "garden." Regardless, our interview question was easily understood by the Portuguese participants in the way that we meant: to refer to open green areas within Lisbon where one can readily encounter grass, plants, flowers, trees, benches, play areas, and so forth. Our results showed that—in contrast to their reasoning about pets—participants often expressed awareness of the physical and psychological benefits that parks afforded:

Green spaces give us clean air and that could disappear. For instance, there in Parque Monsanto, they are the lungs of the city. If we cut [the trees] *down, we wouldn't have that anymore.*

[Gardens are important] *because the city is a place that causes great stress and it gives a chance to someone to go to a place that is near, and to be in contact with nature, to stay calm.*

When we examined the reasoning of the participants who said that parks were important (100%, noted above), we found that 19% of their justifications appealed to physical welfare and 12% to psychological welfare.

Environmental Problems and Solutions

Virtually all the participants (96%) were aware of environmental problems. Out of the total number of environmental problems mentioned (270), participants most frequently mentioned problems of pollution (47%), including pollution to the air and water, garbage, and too much noise. Then, in decreasing order, participants mentioned problems concerning harm to animals (15%), the ozone (14%), urban development (5%), and nuclear energy/weapons (4%). Only one participant mentioned overpopulation as an environmental problem, although multiple responses were encouraged. This result is surprising because many researchers and environmentalists argue that overpopulation is perhaps the most fundamental and pressing problem currently facing our planet (Daily and Ehrlich 1997–98; Irvine 1997–98; McKibben, 1998; Mills, 1997–98).

When questioned directly, virtually all of the participants said that air pollution (98%) and too much logging (97%) constituted environmental problems within their country. For both categories of problems we then asked: "If you were the ruler of the world, what would you do to solve this problem?" In this way, we sought to understand how participants would approach solving environmental problems if they were empowered politically.

In our analyses of their proposed solutions to the problems of air pollution and logging, five types of measures emerged: prohibitive, affirmative, technological, compensatory, and transformative. *Prohibitive* measures sought to curtail or prohibit certain actions ("I would say that each family could have just one car"). *Affirmative* measures sought to implement proactive policies ("subsidize the farmers who many times are peasants with very little to live by [so that] their pine trees [have a] longer time and let them grow"). *Technological* measures sought to promote the creation of new technologies or to promote the distribution of existing technologies ("They should have treatment centers like in France, where they treat the trash before it goes into the rivers"). *Compensatory* measures sought to balance harmful activity with helpful activity ("I would impose certain criteria of rationality—that is, each tree that is cut down, one has to plant a new tree, so nobody would cut too many and it would compensate"). *Transformative* measures sought to change people's beliefs, attitudes, and values ("Everything comes from the fact that you have to change people's personality—to prohibit or to impose fines is not the way that is going to cause people to change their ways of thinking").

Participants offered such measures with the following frequency (for solving problems related to air pollution and logging, respectively): prohibitive (39% and 42%), affirmative (22% and 16%), technological (26% and 42%), compensatory (0% and 26%), and transformative (13% and 12%,). Interestingly, this last measure maps well onto Aldo Leopold's view of environmental education. The reader may recall the passage I quoted in the Introduction where Leopold ([1949] 1970) writes of his disappointment with the slow progress in conservation education: "No important change in ethics was ever accomplished without an internal change in our intellectual emphasis, loyalties, affections, and convictions" (p. 246). In chapter 12, I will build on such a perspective by offering a constructivist account of environmental education.

The Case of the Polluted Waterway

In this first scenario—"The Case of the Polluted Waterway"—we paralleled our earlier pollution scenarios of past studies so as to have a clear basis for comparison. Results showed that all participants (100%) judged the individual act of throwing garbage in the Rio Tejo as not all right. Participants maintained their judgments about not throwing garbage in the river even in conditions where local conventions legitimated the practice for their entire community (100%) and for a community in a different geographical location, along the Amazon River in Brazil (95%). Basing an assessment of moral obligation on negative evaluations across all three evaluations, results showed that 95% of the participants viewed polluting the Rio Tejo as a violation of a moral obligation. Moreover, 99% of these participants used moral justifications in supporting either their prescriptive judgment or their judgment that common practice does not legitimate the act. These results replicate the findings from the Houston child study, the Brazilian Amazon study, and the Prince William Sound study. Thus we now have support for the proposition that moral obligation can underlie not only children's but adolescents' and young adults' environmental reasoning.

The Case of the Driven Automobile

In this second scenario—"The Case of the Driven Automobile"—we set up a situation in which participants judged the moral status of driving cars. We proceeded as follows: First, we asked whether air pollution was a problem in Lisbon. As noted earlier, 98% of the participants said yes. Then we asked whether driving a car increases the air pollution. One hundred percent of the participants said yes. In this context, we then asked a pivotal question: "Do you think it is all right or not all right that a person drives his or her own car to work every day?" For those participants who said it was all right, we then countered with the probe: "But how is it all right to drive the car if, as you said before, that increases the air pollution?" For those participants who said it was not all right, we then countered with the probe: "But how could this person arrive at his or her place of work? Would that be practical?" Results showed that, in some form or another, 81% of the participants believed it was all right for a person to drive his or her car to work. But the reason I say "in some form or another" is that in various ways participants often qualified their evaluations and sought to coordinate their judgments about pollution with other personal and moral considerations of import. Thus what we had expected would be a straightforward assessment of an evaluation turned into a more complex analysis of the coordination of judgments.

Let me then back up and remind the reader of where we stand with the idea of coordination analyses. In chapter 4, I proposed that some of the confusion in the moral-developmental literature has arisen because of inattention to the ways that people can and often do maintain multiple judgments. For example, consider the familiar proposition: "People have the right to free speech, but they don't have the right to shout fire in a crowded theater." That proposition can be coherently maintained because in specific contexts the right to free speech can come into conflict with, and be overridden by, other moral considerations, such as not inciting a stampede in a movie theater. In chapter 4, I also showed that moral considerations can conflict not only with other moral considerations but those that are personal and conventional (Smetana 1983; Turiel and Smetana 1984). Accordingly, at that time I described my programmatic research strategy: that as an initial thrust I would investigate environmental moral reasoning in largely unconflicted situations. My expectation—which I believe has been met—was that such a strategy would (a) provide clarity on how moral environmental ideas are prototypically structured, and (b) help uncover otherwise elusive environmental views and values.

Within this context, some coordination analyses have already been offered, as would perhaps be inevitable looking at any set of complex social-cognitive data. For example, in the Prince William Sound study I showed how biocentric reasoning arises through the coordination of human-oriented and nature-oriented considerations. Moreover, in the previous studies (and this current one), we assessed moral obligation in terms of a particular form of a coordination whereby a prescriptive judgment (that an act is not all right to perform) is not overridden by legal or conventional counterclaims. What now has become apparent is that discretionary moral reasoning also reflects forms of coordination. Specifically, from "The Case of the Driven Automobile" we were able to ascertain three overarching forms by which participants coordinated their judgments concerning the air pollution caused by driving a car with the permissibility of driving: overriding, contextual, and contradictory.

In an *overriding coordination,* one consideration simply overrides other considerations. It can take the form of obligatory reasoning discussed above. For example, in supporting a judgment that driving a car is fundamentally not justified, one participant said: "I think that is totally not all right. Because I think that in Lisbon there is good public transportation . . . that comes at reasonable frequency and that is not expensive." But it is equally possible to use an overriding coordination to argue that driving a car is fundamentally permissible. For example, another participant said, "I think that it is right. Because one needs this asset to go to work, so he won't have to face long lines, like the ones for the buses, so he won't waste so much time." Although these two evaluations differ, the nature of the coordination is the same insofar as each participant upholds a single generalized position. In a *contradictory coordination,* contradictory positions are upheld. For example, one participant said, "It's right because there are a lot of people who don't have public transportation to go to their jobs. . . . Well, it's a contradiction, but it is that way." This participant said "it's a contradiction" because she had just established that the action in question was not all right. In a *contextual coordination,* the judgment is dependent on the specific context. For example, one participant said "It depends. If the place of work is very far away and there is no other way of transportation, then one has to take [one's car]. But if there are other ways of transportation that cause less pollution, I think that people should go [that way]. . . . One could also go by bicycle, that helps exercise and doesn't cause pollution."

Quantitative results showed that 32% of the participants provided overriding coordinations, 33% contradictory coordinations, and 35% contextual coordinations. Also, I can now be precise about what I meant earlier when I said that, in some form or another, 81% of the participants believed it was all right for a person to drive his or her car to work. Specifically, this number includes affirmative overriding coordinations, contradictory coordinations, and contextual coordinations.

Almost tautologically, little amounts of pollution do little harm; and if many people create little amounts of pollution, those little amounts can add up to large amounts that cause significant harm. Accordingly, we asked participants a similar evaluative question concerning the permissibility of driving, not for an individual but (a) for their entire community, and (b) for a community in a different geographical location, New York City. Based on the same type of coordination analyses described above, about half of the participants said it was not all right for the majority of people to drive their cars to work in Lisbon (54%) and New York City (54%). In addition, we asked participants whether they thought it would be better if people did not drive their cars to work. At stake here is whether an act would be considered morally good even if it is not required. Results showed that 89% of the participants said it would be better if a single person in Lisbon did not drive his or her car to work; 86% of the participants said it would be better if everybody in Lisbon did not drive their cars to work; and 89% of the participants said it would be better if everybody in New York City did not drive their cars to work. Thus, depending on the participant, "The Case of the Driven Automobile" pulled sometimes for obligatory moral reasoning but more frequently for discretionary moral reasoning—where the individual act (of driving a car) was viewed as morally permissible, but better if not performed.

Environmental Moral Justifications

As shown in table 10.1, many of the justification categories employed in this study replicate those used in our previous studies. These categories need little introduction. There is, however, one modification in particular that I would like to highlight before examining the justification data quantitatively.

Isomorphic and Transmorphic Biocentric Reasoning

In the previous studies I have shown how many forms of biocentric justice reasoning establish moral standing for nature vis-à-vis moral standing for people. Recall Arnold, for example, from the Houston child study, who said, "Fishes, they want to live freely, just like we live freely. . . . They have to live in freedom, because they don't like living in an environment where there is much pollution that they die every day." In characterizing this reasoning in chapter 6, I said that Arnold viewed an animal's desire for freedom ("to live freely") to be equivalent to that of a human's desire, and because of this direct equivalency animals merit the same moral consideration as do humans. In the current study, we saw a good deal more of such justifications, and with enough detail so as to be able to better cast our characterizations in terms of what we call isomorphic and transmorphic reasoning.

In *isomorphic* reasoning there is an appeal (as in Arnold's case above) that is based on recognizing value or justice correspondences between humans and other natural biological or nonbiological entities. Here is an example of isomorphic justice reasoning from the Portugal study:

Table 10.1
Summary of environmental justification categories

Category	Summary description
Anthropocentric	An appeal to how affecting the environment affects human beings.
Personal	An appeal to personal predilections ("because I love fish"), personal interests ("because if the Rio Tejo were clean, we could swim in it"), or personal projects ("people get to know each other in the gardens").
Relational	An appeal to a relationship between humans and nature, including an appeal to companionship ("[plants] are important because as with the animals they keep us company") or to taking care of aspects of nature as one might take care of a person ("because we can give love to animals").
Welfare	An appeal to the physical, material, and psychological welfare of human beings, including the self, other individuals, individuals within a larger systemic social context or ecological context, or future generations ("I would [care if the water were affected because] look, again, it is a very selfish theory . . . From an economic point of view the water would be captured and sent to a central plant where it would be treated. Who is paying for the process to clean the water? Isn't it us? So, we are causing harm to ourselves.")
Justice	An appeal that humans have rights, deserve respect, fair treatment, or ownership of property, or merit freedom ("because it is polluting the water . . . and nobody has the right to make it dirty, it belongs to the public").
Aesthetics	An appeal to the preservation of the environment for the viewing or, more broadly, sensorial pleasure of humans ("because dirty water is unpleasant, there is no comparison to see a river with clean water, to see the fish swimming, to see the pebbles, and to see that brown, grayish, thick disgusting water").
Biocentric	An appeal to the moral standing of an ecological community of which humans may be a part.
Intrinsic value of nature	An appeal that nature has value, including a focus on biological life ("[wild animals are important because] every living being has to have the opportunity to be alive"), natural processes ("[wild animals] are important because they maintain the balance of the ecosystem"), or telos of nature ("[wild animals] are important because if someone created them it is because they have some kind of role"), including appeals established by means of isomorphic

Table 10.1 (continued)

	and transmorphic reasoning ("they [plants] are important, as the animals are important, because they are living beings and live like us").
Harmony	An appeal to a conception of harmony between humans and nature ("because it is not going to be in harmony . . . there will be a lack of balance").
Justice	An appeal that nature has rights, deserves respect or fair treatment, or merits freedom ("[wild animals are important] because I think that all animals have the right to their life"), including appeals established by means of isomorphic and transmorphic reasoning ("because I think that in the same way that we procreate, they also have the right to live, to be happy . . . because I think that they were also created the same way that we were, and because we have the right to live, everybody has a right to live").
Harm to nature	Although no reference is made to whether appeals for nature derive from an anthropocentric or biocentric orientation, such appeals include a focus on animals, vegetation, nonliving parts of nature, species, natural process, food chains, or ecosystems ("I think it is wrong [if one person throws their trash in the Rio Tejo because] it is like helping to pollute the river, and not only the river, it is also the ground").

It would matter to me [if birds were harmed] *because I think that the animals have as much right to live and to have good conditions of life as we do, and the pollution that affects us will affect them also.*

What the term isomorphic captures is the idea of some form of a symmetrical correspondence between humans and nature. In turn, *transmorphic* reasoning takes an isomorphism and then extends it either through compensatory or hypothetical considerations. For example, one participant said:

[Wild animals are important] *because they breathe like we do, and sometimes we think that because they are animals they are not like us, that they don't do certain things. Then we end up seeing that they do.*

Here this participant understands that animals are in certain respects different from humans ("they don't do certain things") but also similar ("they breathe like we do"), and that such differences do not void that mapping of similar value considerations from humans to nature.

Let me be clear that in comparison to our previous studies we are not so much providing different data but, rather, more specificity. In addition, we are offering what we take to be a more elegant label (and overarching conceptualization) that should be more inclusive for characterizing new data in future studies.

Quantitative Results

The quantitative results are reported in table 10.2. Two overall findings emerged. The first is that the eleven questions pulled more for anthropocentric than biocentric reasoning. Only two questions elicited more than 30% of biocentric justifications—why wild animals are important, where 73% of the justifications were biocentric; and why participants would care if the birds were harmed, where 34% were biocentric. The second overall finding (not shown in the table) is that there were very few developmental results. On occasion, within subcategories, some trends emerged. For example, the intrinsic value justification was used with the following frequency in explaining why wild animals are important: fifth-grade (43%), eighth-grade (67%), eleventh-grade (60%), and college (71%). But such trends were not pervasive.

Taken together these results offer qualified support for the proposition I had advanced in earlier chapters, wherein I suggested that biocentric reasoning might emerge more fully in adolescents and adults. For the majority of participants in this study, it appears that biocentric reasoning has taken shape structurally. But such reasoning manifested differentially and somewhat infrequently across a range of environmental issues.

Conceptions of Harmony

As in the Houston parent study and Prince William Sound study, we sought to understand participants' conceptions of harmony. We pursued this issue by asking three related questions: First, we asked, "For you, what does it mean to live in harmony with nature?" Second, we asked, "How do you know if someone is living in harmony with nature?" Third, we asked, "Can you give an example of someone living in harmony with nature?" Table 10.3 presents the summary of the coding categories that emerged by analyzing all three questions together.

Table 10.2
Percentages of environmental justifications by question and category

Justification category	Pets important	Wild animals important	Plants important	Parks important	Act evaluation (water)	Contingent evaluation (water)	Care about fish	Care about birds	Care about water	Care about the view	Care about people
Anthropocentric											
Personal	12	2	1	29	5	7	5	8	10	11	2
Relational	69	1	1	0	0	0	2	2	0	1	0
Welfare	15	17	49	35	18	14	43	18	57	11	46
Justice	0	1	0	1	6	6	1	1	1	0	29
Aesthetics	1	4	29	31	13	11	15	25	19	73	18
Biocentric											
Intrinsic value	2	59	16	2	1	1	4	17	4	1	2
Harmony	0	5	1	0	2	1	2	0	0	0	0
Justice	1	9	1	0	2	2	15	17	1	0	1
Harm to nature	0	2	1	3	53	59	14	12	8	3	1

Source: Author's calculations.
Notes: Percentages may not equal 100 because of rounding. Multiple responses were coded.

Largely, the data that comprise these categories are not so very different from those of our earlier studies. But, as with our justification coding system, we changed some of the category characterizations to achieve greater fidelity to the data, and we changed some of the category labels in ways that we think are more elegant. Specifically, we combined the earlier categories of "possessing" and "acting upon" into the category "physical." We recast "experiencing" and "state of mind" in terms of "sensorial" and "experiential." And we combined "respect for nature" and "balance with nature" into the category "compositional." To be clear, by compositional we do not mean something like "additive com-position," where there is a stepwise addition to a structure. Rather, we mean composition as in a musical or artistic composition, where one seeks an overarching integrity, beauty, sense of balance, or proportion, and where one focuses on the entire entity and the ways in which the pieces support the whole. Thus our previous category of "balance with nature" comprises but one form of compositional reasoning.

Based on these revised categories, our quantitative results showed the following pattern of usage: physical (27%), sensorial (3%), experiential (5%), relational (24%), and compositional (41%). Developmentally, compositional reasoning was virtually unused by the fifth-graders, and increased with age: fifth-grade (3%), eighth-grade (31%), eleventh-grade (52%), and college (71%). This developmental result largely agrees with what we found in the Prince William Sound study, although the similar orientation ("balance with nature") emerged a little earlier in terms of grade level—with the fifth- and eighth-graders (compared to the second-graders).

Conceptions of the Natural

During the summer that preceded the time of interviewing, many forest fires had erupted in Portugal, due both to natural causes (such as light-ning) and unnatural causes (such as arson). This situation provided an ideal context in which we could investigate how participants conceived of the natural, and of whether human activity counted as natural.

We began this investigation by first assessing whether participants believed the forest fires were natural. Results showed that 3% said

Table 10.3
Conceptions of harmony—summary of categories

Physical	Conception based on doing something to nature, for nature, or with nature, including *negative acts* ("Harmony with nature is not to destroy trees, not to destroy nature"), *positive acts* ("Harmony means to protect the animals and the plants"), and *activity* ("When a person is living in harmony with nature he goes to the countryside and has a picnic").
Sensorial	Conception based on apprehending nature directly with the senses ("Harmony means seeing everything blooming, not seeing people cutting trees down, smelling nature's environment").
Experiential	Conception based on experiencing a particular state of mind or feeling ("Harmony means feeling comfortable with yourself in that moment and in that place").
Relational	Conception based on a relationship between humans and nature, including *personal caretaking* ("[Harmony means] when I see a wounded animal, I help it") and *psychological rapport* ("[Harmony means] talking with the trees. . . . Sometimes I talk to them as if they were people, like this").
Compositional	Conception based on being in balance with nature, including a focus on *anthropocentric compositions* ("We can live in harmony with nature without having to destroy more than we are allowed; nature has *x* resources to give us, and if we take them all at once, we leave nothing to grow") and *biocentric compositions* ("To live in harmony, it is the balance, we trade with nature in a way that none of the parts suffer any harm").

natural, 42% said unnatural, and 55% said it depended upon the fire. Then, upon further questioning, we found the following: 97% of the participants said a forest fire is natural if caused by lightning, 99% of the participants said a forest fire is not natural if a person caused it on purpose, and 90% of the participants said a forest fire was not natural if a person caused it accidentally. Thus these results suggest that even when factoring out intentionality, the participants still viewed humans as separate from the natural world.

We pursued this last issue more directly by asking participants, "What does it mean to say that something is natural?" Results showed that in their conceptions of the natural, 94% of the participants viewed humans as apart from nature. In their reasoning, participants often employed either a negation (35%) or an affirmation (59%). In a negation, the natural was understood as that which remains after one has factored out the human component:

It is natural when there is no human intervention, when it is not our intention to cause that event.

Something is natural when it is not made [by a person] *. . . without us having to do anything.*

In an affirmation, participants affirmed the spontaneous qualities of nature:

[Natural] *means that it comes from Nature . . . came up spontaneously because of excessive heat, or because the wind blew some dust, a spark.*

[Natural] *means that something is spontaneous, that happens or comes from nature.*

I should mention that such affirmations often coexisted with negations ("[the natural is] something that happens spontaneously, without man's intervention"), but we coded such reasons as affirmations because the latter appears to us as the more complex form of reasoning.

I said above that in conveying their conceptions of the natural, 94% of the participants viewed humans as apart from nature. In turn, our results showed that only 3% of the participants viewed humans as a part of nature. Here are two examples:

It is natural [for people to cut down the trees] *because we've always done that. . . . I think that it is nature, too, because it comes from ages before, it is something rooted in people,* [it] *is like the elephants that put trees down, too.*

It is natural [to cut down the trees in the forest because] *we have to take advantage of the resources where we live and where we have them. . . . I think that nothing is more natural than to take advantage of what is given to us, of what we can do with those things.*

Moreover, only 3% of the participants sought to integrate the idea that people exist both apart from and as a part of nature, if only by recognizing the resulting conundrum:

That is a hard question to answer [what does it mean to say that something is natural?]. *Because we are the natural causes, we are part of nature, we are part of the environment. I am contradicting myself* [just before she said that a fire isn't natural if it's caused by a person]. *If we were part of nature and of the environment, the things we cause are natural. That is why I am contradicting myself. What are natural causes?*

Linguistically, it is not unreasonable to hold the position that humans are fundamentally separated from the natural world. After all, if the natural includes everything that is human and nonhuman, then it includes everything; and as Rolston (1997) points out, one "cannot refer to everything and get any meaningful work done with words" (p. 41). But the problem with separating humans from the natural is that this position fails to account for the obvious: that we are biological beings with an evolutionary history and subject to natural laws. In such ways, we are as natural as anything else. Given the seeming persuasiveness of both positions, it was surprising that participants overwhelmingly favored the first.

Generational Amnesia

In the Houston child study, I provided tentative evidence for what I called environmental generational amnesia. Children (and particularly the first- and third-graders) understood about water, air, and garbage pollution in general, but did not believe their own city was polluted in these three ways. I suggested that for children to construct understandings of pollution they need to experience places that are less polluted; otherwise to them their amount of pollution is not pollution but simply the norm. Similarly, I suggested that from generation to generation the amount of environmental degradation increases, but each generation in its youth is inclined to take its amount of pollution as the norm, as the nonpolluted state.

The Portugal data extends our analysis here, helping us to see pathways by which children, adolescents, and young adults learn of their own city's pollution. Sometimes these pathways appeared to follow the proposition suggested above, as participants compared their environment with a direct experience of a pristine area:

My grandmother lives in the North and I go there. And there are many rivers that still aren't polluted. And I think that, I go up there and then I come back, I see up there a river that is not polluted. I feel the water running and I come back down here, I see trash, I think that there is such a difference. And I would like that the Rio Tejo—because I live in Lisbon, I was born in Lisbon—would like that the river in my hometown were not so polluted.

More often, however, it appeared that participants made such comparisons based on culturally conveyed knowledge:

Do you think that throwing trash in the Tejo affects the view and the landscape? *I do; actually, it is already affected, right? How? It becomes ugly, it smells bad. I remember, for instance, a person who still talks about the time when he used to swim in the Rio Tejo, and that he misses that a lot. And I, just eighteen years old, find it difficult to believe that this was possible. However, that was the main source of enjoyment of that person.*

[It would matter to me if the Rio Tejo was affected] *because I heard that some time ago, when there was none of that pollution, the river was, according to what I heard, was pretty, there were dolphins and all swimming in it. I think it should have been pretty to see, anyone would like to see it.*

Thus these latter comparisons provide instances of how culture provides information with which people construct ideas, and highlight the importance of environmental education.

Gender and the Human Relationship with Nature

In a review of the literature on concern for the natural environment, Chawla (1988) writes: "One of the salient findings of survey research is that men tend to have greater knowledge about the natural world, but women tend to express greater concern for it" (p. 17). Kellert (1996, chap. 3), for example, found that compared to men, women revealed greater humanistic and moralistic sentiments to the natural world, more emotional attachments to domesticated animals, and a greater likelihood to join groups opposed to the consumptive use of animals. In turn, in comparison to women, men revealed a greater intellectual orientation to the natural world, more interest and less fear of wildlife and nature, and a greater likelihood to be members of hunting and fishing organizations. Moreover, some of these gender differences have been found to emerge in childhood (Bunting and Cousins 1985; Chawla 1988).

Yet throughout this book I have been remarkably silent about gender differences. The reason is that based on statistical analyses we have found so few—indeed, no more than would occur by chance given the many hundreds of statistical tests we have performed. This same pattern held up in the current study: Virtually no gender differences emerged statistically for evaluations, content responses, or justifications. If, as Mohai (1997) suggests, the effects of gender are modest at best, then it is possible that our comparatively small sample sizes (compared to survey research) have not allowed for enough power in our statistical tests to uncover statistical differences. Regardless, such differences, to the extent they exist, need to be understood within the context of what appear to be substantial structural similarities in cognition and values.

To provide the reader with a sense of what we have been looking at qualitatively, consider five matched pairs of reasoning within justification categories that (based on the research cited above) one might be inclined to view in gender-specific terms: psychological welfare, relational, aesthetics, anthropocentric justice, and biocentric justice. For each pair, I withhold briefly the participant's gender until the subsequent characterization so as to allow the reader a fresh look at each example.

[Gardens are important] *because the city is a place that causes great stress and it gives a chance to someone to go to a place that is near, and to be in contact with nature, to stay calm.*

[Gardens] *are important because in the middle of so much pollution and so many cars and so much stress, they are a way for people to relax.*

Both the male (first example) and female (second example) recognize that the city causes stress ("the city is a place that causes great stress,"; "in the middle of . . . so much stress") and that the public gardens help a person to relax ("to stay calm,"; "to relax").

[Domestic animals are important because] *for the adult who feels lonely it helps to keep him or her company. They are very important to old people.*

[Domestic animals] *are important because when people are lonely, without anybody else, animals can be companions.*

Both the female (first example) and male (second example) focus on the benefits of companionship that domestic animals provide to people who are lonely ("for the adult who feels lonely it helps to keep him or here company";"because when people are lonely . . . animals can be companions").

[It would matter to me if the water was harmed] *because . . . dirty water is unpleasant, there is no comparison to see a river with clean water, to see the fish swimming, to see the pebbles, and to see that brown, grayish, thick, disgusting water.*

[I would worry about how the landscape was affected] *because I think that we all like to see pretty things, things that are pleasant, and the trash in the Tejo is not that at all, things that are pleasant to everybody. I would like to know one person that would say, "Look, I like to watch the trash going by."*

Both the male (first example) and female (second example) appeal to the viewing pleasure of humans ("there is no comparison to see a river with clean water";"we all like to see pretty things"). Indeed, if anything, the female here casts her appeal in a more generalized form ("I would like to know one person that would say, 'Look, I like to watch the trash going by.'")—a trait sometimes attributed more to males than females in the feminist literature.

[It's not all right if everyone in Lisbon threw trash in the Rio Tejo] *because it is polluting the water, and nobody has the right to make it dirty, it belongs to the public. Nobody, nobody, not even a group, not even by oneself.*

It is wrong [for a person to throw trash in the Rio Tejo] *because one has no right to make dirty what belongs to everybody.*

Both the female (first example) and male (second example) view the act of polluting the river as not within a human's rights ("nobody has the right"; "one has no right") because the river is understood to belong to everybody ("it belongs to the public"; "what belongs to everybody").

[Wild animals are important] *because I think that they also have the right to live in the jungle. It is not just us that have to live. Because I think that in the same way that we procreate, they also have the right to live, to be happy.*

[It's not all right that the community in Lisbon threw garbage in the Rio Tejo because] *it would destroy the environment, and we don't have the right to do that, because we are living beings, the same as the others.*

Both the female (first example) and male (second example) appeal to rights ("they also have the right to live"; "we don't have the right to do that") by establishing an isomorphism between animals and humans ("in the same way that we procreate"; "we are living beings the same as the others").

Cross-Cultural Comparisons

Some of the questions in this study paralleled the questions we asked in the Brazilian Amazon study and the Houston child study. In this way, we were able to perform direct cross-cultural comparisons. Table 10.4 presents the quantitative results across all three studies. As shown by this table, by and large participants across all three studies shared similar environmental values and knowledge.

Equally important, as noted earlier, the Portuguese participants' moral obligatory reasoning about the pollution of their local waterway replicates the findings from the Brazilian Amazon study and the Houston child study.

Finally, the qualitative data help sustain the position that common forms of environmental reasoning and values cut across cultures. I have already examined some qualitative commonalities in the above sections. For example, in discussing the environmental justifications I provided an example from the Houston child study and Portugal study to illustrate the idea of a biocentric transmorphism, whereby recognized differences between animals and humans do not void the mapping of similar moral considerations from humans to animals. In addition, below are four more matched pairs. I present them because universality plays a central role in my account of the human relationship with nature and I should like the reader to get as close as possible to the interview data so as to take a look at what my colleagues and I have been seeing.

[T]*he smell of the water, it should bother people to open their windows and feel that foul smell.* . . . [It would matter to me] *because a person shouldn't have to smell dead fish or trash bags full of rotten stuff when she opens the window in the morning.* (Portugal study)

[The air] *stinks, 'cause I laid up in the bed the other night. Kept smelling something, knew it wasn't in my house, 'cause I try to keep everything*

Table 10.4

Percentage of participants' environmental values and knowledge

Environmental criteria	Portugal study (n = 120)	Brazilian Amazon study (n = 44)	Houston child study (n = 72)
Animals important	96	100	84
Plants important	97	98	87
Parks, gardens, and open spaces important	100	—	70
Aware of environmental problems affecting self or community	96	83	78
Discuss environmental issues with others	79	63	72
Act to help solve environmental problems	90	54	86
Thinks that throwing garbage in a river harms birds	91	93	94
Cares that birds would be harmed	95	98	89
Thinks that throwing garbage in a river harms the water	100	—	95
Cares that the water would be harmed	99	—	91
Thinks that throwing garbage in a river harms the view	98	98	92
Cares that the view would be harmed	97	93	93
Thinks that throwing garbage in a river harms the people along the river	100	95	91
Cares that the people would be harmed	96	88	83

Source: Author's calculations.

Notes: a. Participants were first asked if they thought harm occurred (to the river, water, view, or people). Only those participants who thought harm did occur were then asked if they cared about the harm. b. The dash indicates that a comparable question was not asked of that group. c. In the Brazilian Amazon study and Houston child study, the questions concerning animals, plants, and parks were framed in terms of whether that environmental criterion was important in the partcipant's own life; in the Portugal study the comparable questions were framed in terms of whether that environmental criterion was important (without specificity to whom or what). d. In assessing whether animals were important in the Portugal population, questions were asked in terms of domestic and wild animals; in turn, we required an affirmative response to both categories to count as an affirmative response to this more general question that animals were important.

clean. Went to the window and it almost knocked me out. The scent was coming from outdoors into the inside and I didn't know where it was coming from. . . . Now, who'd want to walk around smelling that all the time? (Houston parent study)

In the Houston parent study, we examined this parent's response and her universalized objection to pollution ("Now, who'd want to walk around smelling that all the time?"). This Portuguese college student gives voice to the same idea ("a person shouldn't have to smell dead fish or trash bags full of rotten stuff"). Indeed, based on Mohai and Bryant's (1996) review of the literature on race and concern for environmental quality, it would appear that pollution offers one of the most direct experiences that people have of a degraded environment, and that people everywhere who recognize such pollution can be expected to object to it.

I live in the country and I find that living in the city is very difficult. It causes stress. For instance, we live on this street full of trees. Anytime that I leave home in the morning, I feel invigorated seeing the trees and their shade. I can breathe. I can hear the birds. Now, if I lived on a street close to Avenida da Republica, I would feel stressed seeing that amount of cars, very few trees. (Portugal study)

Yesterday, as my son and I were walking to the store and we were walking down Alabama [Street], and for some reason, I think they're getting ready to widen the street. And it's a section of Alabama that I thought was so beautiful because of the trees, and they've cut down all the trees. And you know it hurts me every time I walk that way, and I hadn't realized that my son had paid attention to it, too. (Houston parent study)

In the Houston parent study, we examined this parent's response and discussed her sensitivities to natural vegetation within her community ("it's a section of Alabama that I thought was so beautiful because of the trees"). This Portuguese college student expresses similar sensibilities ("I feel invigorated seeing the trees").

[W]ithout the green areas the planet is not complete. (Portugal study)

[W]ithout any animals the world is like incomplete, it's like a paper that's not finished. (Prince William Sound study)

In the Prince William Sound study, I had offered the second response as an example of teleological reasoning, an appeal to an ultimate purpose or design to the natural world ("[W]ithout any animals the world is like incomplete)." This fifth-grade Portuguese student expresses the same idea ("[W]ithout the green areas the planet is not complete").

[Harmony means] *there has to be a balance between the number of trees that are cut down and the trees that are being planted so there will be a constant number of trees.* (Portugal study)

[Harmony means] *to keep the scales balanced...taking from nature small amounts, but putting back what you take.* (Prince William Sound study)

Here, both participants understand harmony in terms of establishing a balance with nature through a direct compensation ("there has to be a balance"; "to keep the scales balanced").

Conclusion

In the opening to *Paradise Lost,* Milton ([1674] 1978) writes:

Of Man's First Disobedience, and the Fruit
Of that Forbidden Tree, whose mortal taste
Brought Death into the World, and all our woe,
With loss of Eden. . . .

It is Western civilization's creation story, told and retold through the centuries: Our original ancestors gained knowledge but lost their innocence and were cast out from the Garden, separated from the natural world.

Against this backdrop, we examined Portuguese students' conceptions of the natural, from fifth-grade through college. We found that these participants sometimes separated humans from the natural world. For example, in their conceptions of the natural, participants either negated the human and/or affirmed spontaneous (nonhuman) causes ("something that happens spontaneously, without man's intervention"). Moreover, this conception held even when we factored out human intentionality by countering with a situation where a human starts a forest fire by accident.

This characterization—that participants separated the human from the natural—needs, however, to be tempered by our many other results. For example, consistent with our previous studies, we found that the participants valued domestic and wild animals, plants, and parks. Participants were aware of environmental problems, discussed environmental issues with family or friends, and acted to solve environmental problems. Participants believed that throwing garbage in the Rio Tejo would harm fish, birds, water, the view, and people; and participants cared that such harm would occur to each of these categories. Participants' justifications for their evaluations included both anthropocentric appeals (to personal interests, relationships, welfare, justice, and aesthetics) and biocentric appeals (to the intrinsic value of nature, harmony, and justice). In addition, participants' conceptions of harmony cut across five categories: physical, sensorial, experiential, relational, and compositional. Thus all of these findings speak in various ways and to varying degrees of an intimacy that these participants have with the natural world.

In chapter 4, I discussed two overarching forms of moral judgment: obligatory and discretionary. Obligatory moral judgments refer to prescriptive judgments that are also universalized, not contingent on rules, laws, and conventions, and justified based on moral considerations of justice or welfare. In contrast, discretionary moral judgments are those where moral action, although not required of an agent, is nevertheless conceived of as morally worthy based on considerations of welfare or virtue. I had then proposed that both types of judgments would be central to people's environmental reasoning. Yet one of the surprising findings from the previous studies was that consistently we found evidence for moral obligatory but not discretionary environmental reasoning. Why was that? The answer, I believe, is that we had assessed criterion judgments for only one type of environmental stimulus: the intentional pollution of a waterway. At the time, I thought (mistakenly) that the small addition of garbage to an already polluted waterway would, depending on the child, draw out either obligatory or discretionary moral reasoning. I also thought (correctly) that it was important during our early studies to hold a key moral stimulus constant so as to be able to compare our cross-cultural results directly.

In this current study, we managed to achieve both goals by employing two scenarios. The first scenario—"The Case of the Polluted Waterway"—paralleled our earlier pollution scenarios in past studies. Our results provided support for the proposition that moral obligation underlies not only children's but adolescents' and young adults' environmental reasoning. The second scenario—"The Case of the Driven Automobile"—moved us forward insofar as we found evidence that children through young adults can apply discretionary moral reasoning to environmental content. Specifically, some participants understood that driving a car caused pollution, and that it would be better not to drive a car (and to use public transportation instead), yet still considered it permissible to drive to work. Through "The Case of the Driven Automobile" we also uncovered three overarching forms by which participants coordinated their judgments concerning the air pollution caused by driving a car with the permissibility of driving: overriding, contradictory, and contextual. Thus it would appear that just as moral obligatory reasoning reflects one form of a coordination wherein a prescriptive judgment overrides personal, legal, and conventional counterclaims, so does discretionary moral reasoning reflect a different from of a coordination wherein such counterclaims gain purchase.

Taken together our five studies, spanning three countries, have uncovered what appear to be deep commonalities in the development of the human relationship with nature. In the next chapter, I enlarge upon my account. There I propose that by not paying adequate attention to universal aspects of development in general, and moral and environmental sensibilities in particular, we miss many of the essential ways of being human and underestimate our common humanity.

11

Epistemology, Culture, and the Universal

Postmodern scholars across many fields have charged that traditional (modern) research methods and ways of theorizing are culturally biased and philosophically flawed. Such charges could be leveled my way. For example, throughout this book I have offered results that point to potentially universal features in the human relationship with nature. But postmodernists are skeptical at best about the validity of generalizable results. I have also drawn on theories—structural-developmental theory and to a lesser extent evolutionary biological theory—that assume that nature is real, and that nature constrains and shapes developmental processes and pathways. But postmodernists usually maintain that nature is socially constructed, and no more real than are balls and strikes in the game of baseball (Fish 1996).

Thus at this junction, with my five empirical studies behind us, I should like to be clear about how I understand the nature of research and the epistemic status of nature. In doing so I enter the often heated controversies about postmodernity. I sketch two forms of postmodern theory, insofar as they can be distinguished on the basis of their epistemic claims—deconstruction postmodernism and affirmative postmodernism—and I ask whether we should embrace either form. I answer no.

I recognize that this answer presents the risk of alienating people whose work I admire. For example, in his introduction to a series in environmental studies, Griffin (1992), the series editor, writes that "the growing knowledge of the interdependence of the modern worldview and the militarism, nuclearism, and ecological devastation of the modern world . . . is providing an unprecedented impetus for people to see the evidence for a postmodern worldview and to envisage postmodern ways of relating to each other, the rest of nature, and the cosmos as a whole" (p. 6). Similarly, Orr (1992) writes: "Postmodern education must be life-centered"—"designed to heal, connect, liberate, empower, create, and celebrate" (p. 10). In passing, Thomashow (1995) follows suit, and many others make similar claims. What such authors have to say is often right. But I shall hope to persuade that it is right in spite of and not because of their reliance on postmodern theory, and that such theory leads in directions that virtually none of us want to go.

Finally—since a deconstruction of postmodernity is hardly a satisfactory telos for a constructivist—I show how the research agenda offered in this book allows us to embrace the modern project in principle and to reinvigorate it in application.

Postmodern Theory

Modernity goes back a good ways, and its early forms can be found in the pre-Enlightenment writing of David Hume, Francis Bacon, and Galileo. Hume ([1748] 1961), for example, wrote:

It is universally acknowledged there is a great uniformity among the actions of men, in all nations and ages, and that human nature remains still the same, in its principles and operations. The same motives always produce the same actions. . . . Mankind are so much the same, in all times and places, that history informs us of nothing new or strange. Its chief use is only to discover the constant and universal principles of human nature. (p. 83)

Writers around that time were also "convinced that they were emerging from centuries of darkness and ignorance into a new age enlightened by reason, science, and a respect for humanity . . . People came to assume that through a judicious use of reason and education, unending progress would be possible" (Encarta 1998). Thus modernity, as it has come down to us from the past, emphasizes grand theories that look for transhistorical truths and ethical absolutes, and appeals to hierarchy, progress, development, and reason.

Yet the modern project encountered difficulties. As noted by Harris (1992), some of these difficulties arose due to the failure of the program of the positivists, the cultural diversity revealed by anthropology, and the rise of different cultures to the level of international respectability. In response, the postmodern intellectual movement emerged in such fields as literature, art, architecture, and cultural studies. For example, in biblical studies renewed emphasis was placed on the bible, not as literal truth, but as a literary text open to hermeneutic analyses wherein one allows for multiple, equally cogent interpretations. Such pluralism then permeated the social sciences. As Goodman (1972) writes: "There are very many different equally true descriptions of the world, and their truth is the only standard of their faithfulness. And when we say of them that they all involve conventionalizations, we are saying that no one of these different descriptions is exclusively true, since the others are also true" (p. 30).

Deconstruction Postmodern Theory

Deconstruction postmodernists hold tightly to such relativistic tenets (e.g., Culler 1982; Derrida 1978; Foucault 1980; Morss 1992; Norris 1982; Scholes 1989), claiming, for example, that few if any aspects of either the social world (comprised of cultures, subcultures, and individual people) or the physical world are coherently unified or structured, but are rather held together by fragments of our language. Accordingly, deconstructionists ask that we abandon such modern constructs as truth, objectivity, logic, and even rationality. For what is considered true, objective, logical, or rational in one culture or subculture may not be so considered elsewhere, because such constructs only arise out of and gain meaning through specific socio-historical and cultural contexts. Morality, too, is jettisoned. As Grant (1997) writes, according to deconstructionists, "[u]niversal values are also deemed impossible since there is no ultimate reality but language, which, upon examination, is also undeserving of our acceptance" (p. 117).

Toward assessing deconstructionist theory, it is important to recognize three related forms of internal contradictions within the theory itself. First, deconstructionists argue against theory building, yet themselves advance a theoretical position. Second, deconstructionists seek to deconstruct the tools of logic, reason, and rationality, yet they use those very tools to do so. Third, deconstructionists argue against privileging any position, yet if their theory (that holds that no theory can be true for everyone) holds for everyone, even for the person who mistakenly believes it false, then the theory does what it says cannot be done. It privileges itself. It establishes some basis for truth that transcends its own confines. (For a discussion of these and related issues, see Chandler 1997; Crews 1986, 1989; Hoy 1985; Kahn 1991; Lapsley 1996; Lourenço 1996; Kahn and Lourenço in press; Rosenau 1992; Searle 1983; Turiel 1989; Williams 1985; Wilson 1998).

It has been said, however, that deconstructionists are less interested in putting forth a full-bodied coherent theory, and more interested in providing a theoretical platform by which to empower the disenfranchised and right injustices. But such an offering provides less than is first apparent. To illustrate this point, consider a recent controversy that involved Jacques Derrida, who is often credited with founding deconstruction. In 1987, Derrida provided an interview with a French newspaper in which he "explained Heidegger's enthusiasm for Nazism as an outgrowth of Western metaphysics and engaged in a deconstruction of Nazism and 'non-Nazism' in an attempt to show the 'law of resemblance' between them" (McMillen 1993, p. A8). By some accounts this interview taints Derrida by associating his intellectual roots through Heidegger with Nazism, and by highlighting Derrida's attempt to minimize Nazi immorality. Subsequent to this interview, Richard Wolin edited a book published by Columbia University Press that sought to document Heidegger's intimate involvement with Nazism. In the process, Wolin obtained appropriate legal permission from the French newspaper, which holds the copyright to Derrida's article, to translate and publish the interview in his edited book. However, in granting Wolin permission to use Derrida's article, the newspaper never notified Derrida, and when Derrida came upon the published book, with his interview included, he was outraged. In response, Derrida threatened Columbia University Press with legal action unless they halted any further printing of the volume. As a courtesy, Wolin offered to exclude the Derrida interview from further printings—he only required that he be able to include an additional preface that commented on Derrida's actions. Derrida still objected, and Columbia University Press let the book go out of print after several months.

Within this context, it is interesting to note the language Derrida (1993) uses to argue his case in one of several bitter exchanges between himself and Wolin that appeared in the *New York Review of Books:*

I merely demanded that my interview be withdrawn from any subsequent printings or editions . . . . Do I not have the right to protest when a text of mine is
published without my authorization, in a bad translation, and in what I think is
a bad book? As I have since written to him, Mr. Wolin seems to be more eager
to give lessons in political morality than to try to respect the authors he writes
about and publishes, in a greater hurry to accuse than to understand difficult
texts and thinking. . . . (p. 44)

Derrida here is not being entirely unreasonable. True, he did not hold a
legal claim to his interview. But, still, one can argue that, legality aside,
morality requires that an author's permission be given to include the
author's interview in a volume that casts him unfavorably. But Derrida's
own theory of deconstruction seems to disallow the very claims he wants
to make. Specifically, how is it possible for Derrida—who seeks to un-
dermine the very notion of authorship—to claim that he has been mis-
translated? Such a claim would seem to imply that there are criteria or
standards that transcend culture and context by which to judge the merits
of a translation, the antithesis of what deconstruction embodies. Derrida
also asks that Mr. Wolin respect the authors he writes about. But whose
notion of respect are we to respect? Or does Derrida want to suggest that
there is a fundamental core to the idea of "respect for author" that
transcends culture and context? Finally, Derrida talks about his "right to
protest." But are not "rights" part of the baggage of modernity that
Derrida seeks to jettison?

Because deconstructionists propose that no position, idea, or action
can be privileged—that is, judged better, or more adequate, more intellectually sound, more comprehensive, or more moral than something
else—they ultimately have few recourses when injustices occur. True,
deconstructionists can do all the things other people do. In a democratic
society, deconstructionists can write and speak publicly, help draft legis-
lation, use the legal system to press their claims, and run for public office.
But such actions are somewhat disingenuous. The deconstructionist
might run for public office, for example, not because there is a commitment to the democratic process (after all, democracy cannot be privileged
over fascism) but because that is the way within a democratic government
by which to gain power. Power is primary. Power not only subjugates
but liberates. This is the reason that deconstructionists so often emphasize power in their analyses. The added twist to this scenario is that once
deconstructionists gain power, it is very easy for them to fell prey to

perpetrating the same injustices they rebelled against. After all, other groups are the "other," are different, and thus potentially not deserving of the same moral considerations as those of one's own group.

The implications of deconstruction are particularly difficult for many people to accept. Consider, for example, the following description:

[Chinese] guards in Gutsa Prison [in Tibet] raped nuns who were political prisoners and sexually violated them with electric cattle prods. In another prison, the chief administrator said to me, "I will give you Tibetan independence." Then he rammed the cattle prod into my mouth. When I regained consciousness, I found myself in a pool of blood and excrement and I had lost most of my teeth. (Rosenthal 1995, p. A25)

In response to such situations, many people are hesitant to say "live and let live," especially since that very idea is being violated by others. Thus it is my contention that deconstruction as a theory allows for totalitarian political systems, and disregards human rights and dignity.

Deconstruction also disregards nature. As Soule (1995) writes: "[S]ome deconstructionists are taking the next step, claiming that living nature and wilderness are illusory—just some biologists' narrative, banal . . . " (p. 149). In other words, deconstructionists reduce nature to yet another human artifact, invented by language and as easily undone. The implications are as startling as they are troubling: that nature can be trampled because nature does not actually exist (see responses by Hayles 1995, Nabhan 1995, and Rolston 1997); that a virtual reality of nature is more desirable than nature itself, because virtual nature lends itself more readily to manipulation through our language and imagination (see Shepard's 1995 response); and that every culture's environmental ethic is equally valid, and all moral obligations to nature are undone, because morality does not actually exist (see responses by Kellert 1995, and Grant 1997).

Affirmative Postmodern Theory
Many postmodern theorists have been troubled by at least some of the above concerns about deconstruction, in theory and practice. In response, they have attempted to put forth modified positions, which Rosenau (1992) and others have labeled as "affirmative" postmodern theories. Affirmative theories (e.g., Giroux 1988, 1990; Hammer and MacLaren 1991; Hassan 1985; McLaren 1989; Murphy 1988; Richardson 1988;

Weiler and Mitchell 1992; Wyschogrod 1990) still argue for the plurality of value systems but do not maintain that such plurality necessarily leads to the relativism that is so troubling in deconstruction. As noted by Rosenau (1992): "Affirmative post-modernists frequently employ terms such as oppression, exploitation, domination, liberation, freedom, insubordination, and resistance—all of which imply judgment or at least a normative frame of reference in which some definitive preferences are expressed" (p. 136). Moreover, in contrast to the nihilism that often pervades deconstructionist political theory, affirmatives often favor forms of democracy that empower individuals and especially underrepresented groups. At the same time, affirmatives usually embrace a deconstruction-like epistemology that maintains all knowledge is socially constructed.

It is easy to applaud much of the affirmative's agenda. But can affirmatives maintain their nonrelativistic views in light of their deconstruction-like epistemology? Affirmatives think they can, though they are often circumspect in articulating exactly how. As I understand their position, however, the skeleton of their response looks something like this. They maintain that knowledge is not objective. At the same time they maintain that neither is knowledge subjective, because knowledge is grounded in socially constituted relations, bounded by community. As Murphy (1988) says: "[A]narchy is not necessarily the outcome of postmodernism, because public discourse can culminate in the promulgation of social rules" (pp. 181–182). Thus like deconstructionists they deconstruct the objective/subjective polarity, but as affirmatives they maintain that not anything goes: thus postmodernism without relativism.

The problem here lies in believing that majority opinion or community beliefs solve the problem of relativism, when in fact it does little more than raise the problem from an individual to group level. A case in point: Imagine people inside a house without windows are listening to a slight pitter-patter on the roof. After much discussion and factional power struggles, they all agree that it is raining outside. Then a person from outside their community, and literally from outside their house, walks into their house and asserts that it is sunny outside. "A bit windy," she says, "with acorns falling on the roof, but otherwise a glorious sunny day." Now, presumably there are real occurrences of "raining" and "not raining." Presumably in this case the people inside the house are simply mistaken in believing it is raining outside. Thus one can agree that the people inside the house have socially shared knowledge, and that that knowledge goes beyond mere subjectivity of each member. But to say that is not to say that shared knowledge ipso facto validates that knowledge. And the same holds true for ethical knowledge. A community can agree to discriminate against (or torture or slaughter) those outside their community, but such agreements do not establish ethical validity.

Affirmatives might respond by saying that for a community to have valid ethical knowledge not only must members within the community agree to it (thus protecting themselves from oppression), but any time norms are applied to those outside the community, then those outsiders must agree as well. Perhaps affirmatives would thereby establish the following principle: Membership in a democratic community is accorded to those who are affected by its norms, and, in addition, certain norms must protect the minority from majority oppression. A move like this then begins to bound the ethical by establishing universal criteria, and by a conception of what constitutes oppression in a principled and privileged, if not objective, sense. But in so doing, affirmatives begin to embrace a modern epistemology.

To retain the postmodern epistemology, a common response is to make a case based on literary analysis. Surely, it is said, a novel or play lends itself to multiple interpretations. *King Lear,* after all, cannot be reduced to a single meaning. Rather the play's rich and varied tapestry is precisely what allows person after person, generation after generation, to provide fresh and meaningful interpretations. Moreover, if one community says that a piece of fiction is good literature and another community disagrees, is there not room for differing value judgments as well as differing interpretations? If so, then it is claimed that human life itself is rich and varied, like a text, more so, and thus "facts" and statements of "truth" need to give way to multiple interpretations and differing value judgments.

I appreciate the sensitivity that literary analysis can bring to the study of human nature. But life is not literature, and mischief occurs when postmodernists think it so. A deconstruction of physics might provide the "freedom" to generate and offer competing theories about how to understand gravity, but such a deconstruction does not and cannot negate the consequences should we jump off a cliff. Similarly, postmodern architecture can theorize that the "laws" of physics are a cultural or linguistic convention. But try saying that when building a house—in any culture: "One [postmodern] architect is said to have 'built an officers' club, and the roof caved in during the dedication ceremonies.' In other cases postmodern designs are abandoned because 'they simply can't be built' (Seabrook 1991: 127, 129)" (Rosenau 1992, p. 127).

Facts and truths do not stop with physics. Over the last decade, for instance, it has been increasingly clear that child sexual abuse actually happens far more often than previously thought (Bass and Davis 1988; Masson 1984). Many women are not just fantasizing what Freud called "seductions." But think about what is implied from a postmodern epistemology. It would be something like, "You as the woman have your interpretation, and that's important, and it's valid, and you should give voice to it, and become empowered through it." It would also follow, though, that the alleged perpetrator has his own contrary interpretation, and that it is as valid for him as the woman's is for her. In response, the woman could rightly say, "To heck with your theory; the fact is I was sexually abused, raped, and my life is not like literature." This is not to discount the incredible complexity that arises in such remembrances, and that in some cases women may actually remember incorrectly, and unjustly accuse a perpetrator (A Conversation 1994; Tavris 1993). But my claim is that either childhood sexual abuse such as penetration happened or did not, and that one can with validity universalize a judgment that such abuse is morally wrong. As Rosenau (1992) writes: "Modern time, space, and history can be dispensed within post-modern literature, and the results are entertaining. But this is not always the case in the social sciences" (p. 168).

Part of the impetus toward postmodern theory is to appreciate if not celebrate differences in people, cultures, lifestyles, and worldviews. But a theoretical orientation that focuses on more than differences is needed. After all, it would be impossible to understand let alone appreciate another if that other was not—in important and meaningful ways—like us. Imagine if we visit a "strange" people and see them routinely putting organic matter into their mouths and swallowing. We might assume that they, like us, need to eat to survive, and that we are watching people eat. We might be wrong, of course. It is possible, for example, that we are observing a religious rite that has little to do with eating, and that the food in this instance symbolizes something of religious significance. Here, of course, we would be assuming that these people believe in something along religious lines, and that they can and do use symbolic thought. We might be wrong, again. But it seems to me doubtful that human nature can differ too dramatically across such fundamental categories of being (compare Reed 1996). Accordingly, social scientists need a theoretical orientation that can uncover not only differences across cultures, but the ways in which differences are embedded in a larger context of commonality. In what follows, I offer such an orientation, first in terms of moral theory, and then in terms of the human relationship with nature.

Universality (and Differences) in the Moral Life

Anthropological accounts of the practices and beliefs of various cultures provide important data that directly inform on the question of whether the moral life is similar or different between cultures. On a first look, it may seem self-evident that cultures differ morally. For instance, among the practices van der Post ([1958] 1986) documents of the Bushmen of the Kalahari Desert is the abandoning of their elderly either to attack by animals or to sure starvation. From other anthropological accounts we learn that devout Hindus believe it is immoral for a widow to eat fish, or for a menstruating woman to sleep in the same bed with her husband (Shweder, Mahapatra, and Miller 1987). These and hundreds of other practices appear to differ dramatically from the practices of people in Western cultures. But what is crucial in analyzing such anthropological data is to pay close attention to varying moral conceptualizations. Modifying an example used by S. K. Langer ([1937] 1953), consider four men's suits. One is made of cotton, the second wool, the third polyester, and the fourth silk. Each is cut to a different size. Now we ask, are these four objects the same? If by object we mean the material, then the answer is no. If by object we mean their size, then the answer is no. But if by object we mean their function as a suit, then the answer is yes. Thus the answer of whether there are differences or commonalities between the objects depends on our conceptualization.

So, too, with morality. Depending on how morality is conceptualized, and anthropological data is collected, one is led to varying conclusions about moral diversity. For instance, the example of the Bushmen practice of leaving their elderly to die appears fundamentally different from our own practices. But as van der Post further describes the Bushmen's intentions, motivations, social context, and environmental constraints, their practice seems less foreign. The Bushmen are a nomadic people that depend on physical movement for their survival. Elderly people are left behind only when they can no longer keep up the nomadic pace and thereby jeopardize the survival of the entire tribe. When the tribe is thus forced to leave an elderly person behind, they conduct parting ceremonies and ritual dances that convey honor and respect. The tribe also builds the elderly a temporary shelter and provides a few token days of food. All these additional practices convey an attitude of care and concern for the elderly, and felt loss at their impeding death—a death that is unavoidable if the tribe as a whole will survive.

Such an analysis does not negate differences between Bushmen and Western culture. On a behavioral level, both cultures engage in different practices regarding the care of their elderly. But the analysis also points to common ground. Both societies show care and concern for their elderly. Both societies also balance that care with the well-being of society as a whole. Note, for instance, that as some of our medical practices become more extensive and extraordinarily expensive, we face the problem of how to weigh the benefits to elderly patients with the corresponding costs to society. Thus, drawing on the ideas discussed in chapter 4, if we conceptualize morality deontologically, in terms of respect for persons, or consequentially, in terms of promoting the greatest good for the greatest number, then in important respects Bushmen morality may well resemble our own culture's morality.

This type of analysis can be applied to a great deal of anthropological data. For example, Turiel, Killen, and Helwig (1987) reanalyzed the anthropological data of Hindu culture collected by Shweder, Mahapatra, and Miller (1987). In their reanalysis, they found that devout Hindus believed harmful consequences would follow from a widow who ate fish (the act would offend her husband's spirit and cause the widow to suffer greatly) and from a menstruating woman who sleeps in the same bed with her husband (the menstrual blood is believed poisonous and can hurt the husband). While such beliefs differ from those in our culture, the underlying concern for the welfare of others is congruent with our own.

Generally, conceptualizations of morality that entail abstract characterizations of justice and welfare tend to highlight moral universals, while definitions that entail specific behaviors or rigid moral rules tend to highlight moral cross-cultural variation. Theorists who strive to uncover moral universals usually believe they are wrestling with the essence of morality, with its deepest and most meaningful attributes. Recall from chapter 3 the opening to Plato's *Meno* (1956): Menon asks Socrates whether virtue can be taught. Socrates turns the psychological and educational question into a philosophical one, and subsequently asks Menon to explain what virtue is. Menon then defines virtue in different ways, depending on a person's activities, occupation, and age. This answer does not satisfy Socrates:

> If I asked you what a bee really is, and you answered that there are many different kinds of bees, what would you answer me if I asked you then: "Do you say there are many different kinds of bees, differing from each other in being bees more or less? Or do they differ in some other respect, for example in size, or beauty, and so forth?" Tell me, how would you answer that question?

Menon replies: "I should say that they are not different at all one from another in beehood" (p. 30). Which is exactly what Socrates wants to say about virtue, and what we could say about the essence of "suits" in the earlier analogy.

In contrast, theorists who strive for characterizing moral variation usually argue that by the time you have a common moral feature that cuts across cultures, you have so disembodied the idea into an abstract form that it loses virtually all meaning and utility. Reconsider the example of devout Hindus who believe that by eating fish a widow hurts her dead husband's spirit. Is the interesting moral phenomenon that Hindus, like ourselves, are concerned with not causing others harm? Or, as Shweder might argue, is the interesting moral phenomenon that Hindus believe in spirits that can be harmed by earthly activity? In my view, both questions have merit, and a middle ground provides a more sensible and powerful approach in research—one that allows for an analysis of universal moral constructs (such as justice, rights, welfare, and virtue) as well as the ways in which these constructs play out in a particular culture at a particular point in time (compare Asch 1952; Biaggio 1994, 1997; Dunker 1939; Friedman 1997; Hatch 1983; Helwig 1995; Kahn 1994; Killen and Wainryb, in press; Moshman 1995; Nucci 1997; Spiro 1986; Turiel 1998; Turiel, Hildebrandt, and Wainryb 1991; Wainryb 1991, 1993, 1997).

Moreover, it is important to understand that when moral differences do occur between peoples, it is not necessarily the case that the practices of any particular culture are believed legitimate by everyone in that culture. Hatch (1983), for example, reports that women in the Yanomamo tribe in Brazil were "occasionally beaten [by men], shot with barbed arrows, chopped with machetes or axes, and burned with fire-brands" (p. 91). Hatch also reports that the Yanomamo women did not appear to enjoy such physically abusive treatment, and were seen running in apparent fear from such assaults. Psychological data of a similar kind can be found in a recent study by Turiel and Wainryb (1994) on the Druze population in Israel. The Druze largely live in segregated villages, are of Islamic religious orientation, and are organized socially around patriarchal relationships. The father, as well as brothers, uncles, and other male relatives—and eventually a woman's husband—exercise considerable authority over women and girls in the family, and restrict their activities to a large degree. However, when these women were interviewed, "a majority of them (78%) unequivocally stated that the husband's or father's demands and restrictions were unfair" (p. 44). Similarly, consider a true-life narrative of a princess in Saudi Arabia. She writes:

This intimate view of my beloved sister's predicament [that she was married to a 62-year-old man, who tortured her sexually] filled me with a new resolve: It was my thought that we women should have a voice in the final decision on issues that would alter our lives forever. From this time, I began to live, breathe, and plot for the rights of women in my country so that we could live with the dignity and personal fulfillment that are the birthright of men. (Sasson 1992, p. 60)

Thus Yanomamo, Druze, and Saudi women—like many women in Western societies—are often unwilling victims within what they themselves perceive to be an uncaring or unjust society. In such situations, it is less the case that societies differ morally, and more the case that some societies (ours included) are involved explicitly in immoral practices.

Universality (and Differences) in the Human Relationship with Nature

In 1928, Ernst Mayr traveled to the Arfak Mountains of New Guinea to make that area's first thorough collection of birds. Before departing, Mayr studied specimens previously gathered from New Guinea and, based on a Western taxonomic methodology, estimated that he would find a little more than 100 bird species in the Arfak Mountains. Wilson (1992) continues the story:

> Once settled in a camp, after a long and hazardous trek, Mayr hired native hunters to help collect all the birds of the region. As the hunters brought in each specimen, he recorded the name they used in their own classification. In the end he found that the Arfak people recognized 136 bird species, no more, no less, and that their species matched almost perfectly those distinguished by the European museum biologists. (p. 42)

Such similarities between the environmental knowledge of Western and native cultures have begun to be explored formally under the title folk-biology (Coley 1995; Coley, Medin, and Atran 1997; López et al. 1997; Medin and Atran, in press; Medin et al. 1997). In one set of studies, for example, López et al. (1997) used modern experimental psychological methods to compare the taxonomies and resulting inductions between two groups: Itzaj Maya (who inhabit the Petén rainforest region of Guatemala) and Northwestern University undergraduates in the United States. Results showed consensus between both groups in (a) their taxonomies of local mammal species, (b) how their taxonomies differed from a corresponding scientific taxonomy, and (c) their ability to use taxonomies in folkbiological inductions. The results also delineated aspects of cultural variation. For example, the Northwestern University students produced more scientific knowledge and the Itzaj produced more ecological knowledge. Still, such cultural differences did not appear to influence the underlying structure of reasoning of both groups.

A similar conceptualization can be applied to understanding our moral relationship with nature. To show how, I begin with Huebner and Garrod's (1991) claim that Tibetan Buddhism "presents profound challenges to those who argue for general applicability of moral reasoning theories originating in Western culture" (p. 341). They illustrate their point by providing a passage from one of their interviews with a Tibetan monk:

He [the bug] went under my feet, but he did not die. Now he was suffering, wasn't he? Suffering. I figured that if I left him like that, he would suffer forever, because there was no medicine for him as there is for a human being. So I prayed. . . . And then I killed him with my hand, the suffering one. Why did I kill him? He was suffering. If I left him, he would suffer. So it was better for him not to suffer any longer. That's why I killed him. And I prayed . . . that one day in the next life, he would become a man like me, who can understand Buddhism and who will be a great philosopher in Tibet. (p. 345)

Huebner and Garrod say that "such sensitivity to the nonhuman world leads to moral dilemmas not likely considered in Western culture" (p. 345). But is that true? Have not many of us experienced moral qualms very similar to this Buddhist monk—stepping by mistake on ants or caterpillars, or perhaps accidentally driving over a dog or cat and killing it, and feeling remorse? More formally, Western rights-based environmental philosophers routinely trouble over the moral status of animals (Spiegel 1988; Stone 1986). Consider a short passage by Tom Regan (1986), an analytic rights-based philosopher:

There are times, and these are not infrequent, when tears come to my eyes when I see, or read, or hear of the wretched plight of animals in the hands of humans. Their pain, their suffering, their loneliness, their innocence, their death. Anger. Rage. Pity. Sorrow. Disgust. . . . It is our heart, not just our head, that calls for an end, that demands of us that we overcome, for them, the habits and forces behind their systematic oppression. (p. 39)

Regan's sensitivity to the nonhuman world leads him to difficult moral dilemmas. If one accepts that animals feel pain and thereby have moral standing, are people never justified in causing animals harm? How about to advance medical knowledge? Cannot indigenous people justifiably hunt to eat? Cannot we justifiably eat meat?

Such thorny questions, I submit, are considered not just by eminent Western philosophers, but by some of the very Western researchers whom are most often attacked for their blatant disregard for animal welfare. Consider part of a case study of Dr. Dan Ringler, Director for the Unit for Laboratory Animal Medicine at the medical school at the University of Michigan. Ringler directs 70 people, including 12 veterinarians. His group's laboratories house approximately 35,000 animals for their own research (mostly rats, mice, guinea pigs, and hamsters) and 121,000 other animals for other researchers. With my assistance in conceptualization and analyses, Danielle Tarnopol interviewed Ringler about his views toward animal rights and medical research (Tarnopol and Kahn 1996).

From the interview data, it is clear that Ringler, like Regan, grants that animals feel pain. "From a veterinary standpoint," he says, "I think many other animals other than humans have many of the same attributes [as humans]. Our assumption is that pain and any procedure that would cause pain in a human will also cause pain in an animal." In turn, Ringler seeks to minimize the pain in three ways. First, his group is always "looking for alternatives for the use of animals in product safety testing." Second, when animals must be used, his group asks whether "the work can be done in what we call a less sentient animal, one that has a less developed nervous system." Third, "[when appropriate we] give pain relieving drugs where there's aspirin or Tylenol or codeine or morphine or whatever, or anesthetics."

Late in the interview, Ringler says: "I want you to know that in every research institution there are people like us who care about animals, who love animals." If this statement is true, then how can Ringler justify causing animals pain? The answer is as simple as it is pervasive in his reasoning. Ringler starts by rejecting the proposition that animals have rights: "I don't feel that animals have rights. I think we have responsibilities, which come out very close to the same thing." Ringler then moves solidly to a consequentialist moral perspective, wherein he seeks to maximize the benefits to the greatest number of individuals:

I feel we are in the business of reducing suffering on the planet; a lot of pain and suffering worldwide is caused by disease and disability and aging and accidents, and so on. We, medical research wants to learn as much as possible about disease so that we can prevent disease primarily, and treat it if we can't prevent it. . . . From an ethics standpoint I feel that the use of animals in research is appropriate in order to reduce the pain and suffering of both other animals and humans. . . . For example, in the development for distemper vaccine for dogs—it's a horrible disease, it has killed millions of dogs—maybe a thousand dogs were used in development of distemper vaccine for dogs, but since that has been developed and the vaccine has been available millions and millions of dogs have been prevented from suffering the ravages of distemper. . . . Millions and millions of people and some animals benefit from the pain or distress that a few [animals] underwent.

Even more succinctly, Ringler says, "I think it's permissible for an animal to undergo some pain and suffering for the good of the many."

I am not saying that Ringler is correct. Like Regan, I find it difficult to accept that humans can with moral justification intentionally cause animals to suffer. But I am saying that Ringler provides evidence that Western people in perhaps some of the most unlikely places are—like the Buddhist monk—deeply concerned for animal welfare. Moreover, Regan and Ringler together demonstrate how individuals with a similar cultural upbringing (Western, analytic, academic) can share a moral regard for animal welfare but differ in how that concern is manifested normatively and behaviorally. Both also seek to reconcile their moral regard for animal welfare with the potential dilemmas that such regard engenders.

Similarly, across all the developmental studies presented in the earlier chapters, my colleagues and I found evidence that children care about animals, and often faced early forms of the same dilemmas. Consider again, for example, the dialogue with Arnold (a fifth-grade child from the Houston child study) that I used in chapter 6 to highlight the idea of disequilibration and the reorganization and subsequent construction of knowledge. Arnold said, "We really never should kill animals." The interviewer then asks whether Arnold eats meat, and Arnold says, "not that much" and "only when there's rough times and we really need it." Thus there is a bit of a tension in Arnold's reasoning: he first categorically objects to killing animals but then allows for exceptions. Later in the interview, Arnold says:

I love animals. . . . Animals are important to me because I don't like seeing animals being mistreated because every animal needs respect. . . . No matter what life form they're from, no matter how shaped or sized they are.

The interviewer then pushes with another potential dilemma:

Do you have the same feeling about mosquitoes as you do about fish? *Well, not really.* [Laughter] Tell me how that's different? *Because mos-quitoes they begin to get on your nerves a little bit. And they make little bumps on you. But I don't really like mosquitoes. But it's still wrong to kill 'em though. Because they really need to live freely too, just like every insect, every bear, any kind of type of human. . . .*

Thus Arnold faces a dilemma like that of the Buddhist monk: both have sensitivity to the suffering of animals, and both need to find their way in a world where animals, like humans, sometimes suffer tremendously.

Conclusion

I have suggested that postmodern theory, and particularly deconstruction, misconstrues the human relationship with nature by postulating that nature is socially constructed, invented by language, and thereby perpetually malleable. I also suggested that when postmodern theory, and particularly deconstruction, is taken seriously, it leads to contradictions in epistemology, to fragmentation in knowledge, to opportunism in interpersonal relationships, and to nihilism in moral action and commitment. Rather than reject modernity, we need to embrace it and reinvigorate its application. Accordingly, throughout this book I have offered a research agenda on the human relationship with nature that seeks to uncover differences, and give voice to the disenfranchised, while articulating potentially universal features.

12

Environmental Education

One afternoon a hummingbird flew inside our cabin. My wife, daughter, and I were on our land in California, those 670 acres of mountain meadows and forest that I mentioned in the preface. Upon seeing the bird, my then four-year-old daughter, Zoe, followed it, and I followed her, with a plastic container in hand. I trapped the bird against a window, walked outside, and let it go. "Be well and live free," I said. An hour later, I see a butterfly trapped inside. I cup it in my hands and walk onto the porch to find Zoe. When she sees me with my cupped hands she immediately cups hers, and walks up to me. She knows what is happening, as we have done this before on her request. Very gently I transfer the butterfly into her hands. She holds it cupped like that for ten seconds and then opens her hands. The butterfly stays put. Zoe stands poised, quiet, looking at the butterfly on her hands. A minute later the butterfly flies off. Zoe says, "Be well and live free." Later that afternoon Zoe sees a bee drowning in the water. She says, "Dad, quick, get me something to save the bee." I find a lid to a container and give it to her and she dips it in the water. Zoe then positions the lid in different ways until the bee is able to climb on board, and then she sets the lid down on the porch. We both watch the bee. It tries to fly, but cannot. "Dad, it's probably so tired." After a few minutes the bee flies away.

Birds, butterflies, and bees die all the time in nature, and we see our share of such death. I also recognize that from an ecological standpoint it is not necessarily good or even warranted that we save any of them. But as a parent—one seeking to educate his child—I see something else going on during our "animal rescues."

In chapter 8, I asked the ontogenetic question: What comes first, a relationship with people or a relationship with nature? At that time I suggested the answer was neither one nor the other, but both, dialectically: children's moral relationships with other humans help establish their moral relationships with nature, and vice-versa. I had also proposed that this dialectical process initiates the coordinations that lie at the heart of the development of biocentric reasoning. In turn, in many of the empirical chapters, I provided evidence for what such coordinations look like in terms of, for example, justice isomorphisms (e.g., "because I think that [wild animals] were also created the same way that we were, and because we have the right to live, everybody—I think everybody has a right to live").

From this developmental perspective, is it really futile that Zoe and I save a few animals? I do not think so. Through such experiences I see Zoe developing concern for the suffering of other sentient beings, and learning to take delight in their liberation. "Be well and live free." It is the same language I use when Zoe and I talk (in simple terms) about migrant farm workers, slavery during the time of the Civil War, or the Chinese oppression of political dissidents. Indeed, "be well" was one of the themes of the experience I related in the opening to chapter 4, wherein Zoe and I shared a few minutes with a homeless man on the streets of Cambridge one winter day.

Thus in this chapter I take up the topic of environmental education construed broadly, wherein humans are a part of the natural world, and human well-being is counted as an environmental consideration. Given this breadth, and what can be accomplished in a single chapter, I do not seek to provide a comprehensive review of the existing literature on the place and importance of environmental education, and of existing programs, curriculum guides, and published activities (see, e.g., Caduto 1983; Gough 1987; Greenway 1995; Hart 1997; Hungerford 1975; Jinks 1975; Lewis 1981–82; Miller 1981; Orr 1992, 1994; Pomerantz 1986; Rabb 1993; Robottom 1987; Smith 1992; Sobel 1993, 1997; Stevenson 1987; Tanner 1980; Thomashow, 1995; Wigginton 1986). Instead, I take the structural-developmental theoretical framework that has guided my research throughout this book and show how it can make important contributions to the field of environmental education.

Constructivist Education

In chapter 3, I sketched the constructivist position on child development. Briefly, a constructivist believes that development is not adequately explained by (a) endogenous accounts, which emphasize biology, genetics, and evolution, or (b) exogenous accounts, which emphasize behaviorism, social learning, social transmission, and cultural learning. Moreover, constructivists do not believe that development is adequately explained by simply juxtaposing or combining endogeny and exogeny (as in the proposition that development occurs by means of biologically prepared rules of learning). Rather, constructivists place a priority on the processes of assimilation, accommodation, and disequilibration—on the active mental life of children and the ways in which children construct increasingly more adequate ways of understanding and acting upon their world. This constructivist perspective motivated a good deal of the research presented in this book, research through which I sought to understand the child's understanding of nature, and how those understandings are mentally organized (structured) and transformed through development.

Educationally, the implications of this constructivist perspective run counter to many traditional teaching practices and beliefs, which so often and in varied forms embody an exogenous view of learning. To convey this point, and to describe the shift from traditional to constructivist teaching, I elaborate on four constructivist educational principles proposed and practiced by DeVries and her colleagues (DeVries 1987, 1988, 1997; DeVries and Kohlberg 1990; DeVries and Zan 1994).

From Instruction to Construction

One traditional view of teaching is that for students to learn, teachers must instruct, by which it is meant that teachers must correctly sequence curriculum content, drill students on correct performance, correct student mistakes, and then test for student achievement. Granted, one might note a few sidewise embraces of critical thinking and cognition. But if push comes to shove—if, for example, test scores go down—the call is clear: Back to basics. Instruction in the three "R"s. By contrast, the constructivist view holds that learning involves, as I wrote in the introduction, neither simply the replacement of incorrect knowledge with the presumed correct knowledge, nor simply the stacking, like building blocks, of new knowledge on top of old knowledge, but rather transformations of knowledge. Transformations, in turn, occur not through the child's passivity but through active, original thinking. Constructivist education, therefore, centrally involves experimentation and problem solving; and student confusion and mistakes are not antithetical to learning but a basis for it.

From Reinforcement to Interest

Traditional educators often seek to shape student behavior through four types of reinforcement procedures: positive reinforcement, punishment, response cost, and negative reinforcement (Rohwer, Rohwer, and B-Howe 1980). Each procedure presupposes that children learn through stimulus-response conditioning, and that for effective instruction the teacher needs to strengthen, weaken, extinguish, or maintain learned behaviors through such reinforcement procedures. By contrast, as I argued in chapter 2, the constructivist view holds that children do not just process information, which leads to a conception of children as biological machines, but actively make meaning of their world. In turn, children construct meaning more fully when engaged with problems and issues that captivate their interest. Thus constructivist teachers find out what interests their students and then build a curriculum to support and extend those interests. They allow students to help shape the curriculum, and the freedom to explore, take risks, make mistakes. Indeed it can be argued that many of the behavior problems traditional teachers try so hard to suppress—such as students talking or yelling in class, or otherwise not attending to their studies—arise precisely because students find the curriculum drudgery.

From Obedience to Autonomy

In arguing for an exogenous theory of learning, Wynne (1989) says teachers should seek "simple obedience from young persons," and that while teachers may wish to provide reasons for their demands, by and large reasons "merely serve as a form of intellectual courtesy" (p. 2). By contrast, from a constructivist perspective reasons fundamentally help children as they seek moral coherence and understanding in the course of directing their conduct. Thus constructivist teachers move away from demanding obedience and toward fostering the child's autonomy. By autonomy I mean, in part, independence from others. For it is only through being an independent thinker and actor that a person can refrain from being unduly influenced by others (e.g., by neo-Nazis, youth gangs, political movements, and advertising). But by autonomy I do not mean a divisive individualism, as constructivist autonomy is sometimes said to be (Hogan 1975; Shweder 1986). Rather, as I showed in chapter 3, within a constructivist framework autonomy is highly social, developed through reciprocal interactions on a microgenetic level, and evidenced structurally in incorporating and coordinating considerations of self, others, and society. In other words, the social bounds the individual(ism), and vice-versa.

From Coercion to Cooperation

In some sense, the movement from coercion to cooperation reflects another view of obedience to autonomy, but more from the student's than teacher's standpoint. For children to cooperate they must incorporate and coordinate their own feelings, values, and perspectives with those of others. Given that the adult's relationship to children are laden (often necessarily so) with coercive interactions, peer relationships are centrally important to children. Through such peer relationships, concepts of equality, justice, and democracy flourish (Piaget [1932] 1969), and academic learning is advanced (Vygotsky [1934] 1962, 1978).

These four principles help convey the tenor of constructivist teaching, and how it differs from many traditional educational practices. Yet it is also true that constructivist educators can at times employ what may look like traditional practices. For example, although I said reinforcement should give way to interest, it is also true that reinforcement can promote interest. But in such cases the constructivist asks, Whose interests? A parent can give a gold star to his son every time he cleans his room. A teacher can give extra recess time to any student who learns a new set of multiplication tables. Such reinforcement might increase the targeted behavior, but it is more likely the adult and not the child who has interest in that behavior. In contrast, imagine that a child is in the woods, trying to find some bugs to collect and study. The child walks all around, picks up a log, and walks some more, but begins to get discouraged because he cannot find any. An adult could say: "Oh, I really like the way you turned over that log to try to find some bugs. That was a great idea. Let's try it again and see if we can find some." In such ways, positive reinforcement (in the form of verbal praise for a specific behavior) can serve to promote larger constructivist goals.

Or consider again the experience with my daughter, described earlier. I had sometimes initiated the animal rescues in order to "model" behavior for Zoe. Notice also that when the butterfly flew from her hands, Zoe chose words—"Be well and live free"—that mimicked mine exactly. A traditional educator might say: "Zoe just copied you, which is fine, that's what children do, and that's why it's important that parents and educators provide good role models." But the question is, when children model, copy, and imitate are they engaged in largely passive interactions, directly absorbing what their culture presents?

The research literature suggests not. Numerous studies have looked for direct correspondences between the behavior an adult models and the subsequent behavior of a child. For example, in one form of a social-learning experiment (that was particularly common several decades ago), a child is brought into a laboratory setting, and through a game situation with an adult both win some tokens. The adult then models helping behavior by donating some of his tokens to a charity that helps children in poverty. The adult tells the child that he or she is at liberty to donate or not, and then leaves the room. Then an assessment is made of whether or not children donate, and whether such behavior decreases when the adult does not model the behavior. After an extensive review of this literature, Radke-Yarrow, Zahn-Waxler, and Chapman (1983) conclude that although modeling has sometimes been shown to have a direct effect, "the amount of donating after observing a model is often quite low, and not all children adopt the model's behavior . . . and in fact some children act counter to the behavior of the model" (p. 502). Thus from this body of research emerges support for the proposition that children recognize and interpret the complex situational factors that make virtually every social situation unique, and children plan and adjust their conduct accordingly.

More informally, I take this proposition to be in agreement with our general sensibility. Consider, for instance, the effects of television. Most of us have witnessed hundreds if not thousands of television programs that depict and sometimes applaud a protagonist (and thus a potential role model) who burglarizes or murders. But few if any of us have actually committed such acts. Presumably we believe that when we watch such programs we can interpret the depicted acts in ways that prevent us from replicating them: for example, we distinguish what does occur from what we think ought to occur, fact from fantasy, reality from story, and so forth. We also by and large give children credit for such abilities, and consider them to have a significant degree of free will that leads to intentional behavior for which they are held accountable. Thus, not surprisingly, juries have consistently rejected arguments by defense lawyers who have attempted to place the blame of a violent crime not on the young person who committed it, but on the violent television programs that purportedly caused the young person to act violently.

To be clear, I am not saying that exogenous factors do not affect children. Of course they do. They are part of any interactional theory. But I am arguing that the developmental mechanism is not that of a direct adult or cultural installation or transmission of knowledge and norms.

From this perspective, consider a constructivist (as opposed to exogenous) characterization of Zoe's experience while rescuing the bee. First, notice that Zoe chooses a different animal (a bee) to rescue than the two I had (a hummingbird and butterfly). I think she did so because a hummingbird and (at least in the past) a butterfly had been literally beyond her ability to catch. Here with the trapped bee she recognizes an opportunity for more autonomous action. She then directs me to the task at hand: "Dad, quick, get me something to save the bee." Here, too, we see her generalizing from what she had observed (my catching other insects), integrating that knowledge with past knowledge (that she's been stung before by bees), and thus generating a new means/end relation: she needs some implement to free the bee without endangering her hands. I get her such an implement (the lid). Zoe then experiments with the implement until the bee is able to climb on board. Zoe then hypothesizes (incorrectly) why the bee doesn't fly away immediately ("Dad, it's probably so tired"). In summary, from a constructivist perspective I would describe this episode in the following way: Zoe observed what I had done in terms of rescuing animals, made its meaning her own, reconfigured it in a new situation through which she could act more autonomously, cooperatively directed my action, acted upon the resulting construction, hypothesized a cause for an animal's (in)action, and then took pride in her accomplishment. Hardly a passive interaction.

The Difficulty of Grasping Constructivist Theory and Practice

Many people have difficulty grasping constructivist educational theory and practice. To illustrate some of the common difficulties, and ways around them, let me recount a few experiences.

One took place while I was teaching an introductory graduate research seminar. We had covered some of the topics of this book, including structural-developmental theory, some of my own research, and constructivist educational principles. Then I asked students in groups of four to lead our seminar discussions on the biophilia hypothesis. One group, while presenting on Ulrich (1993), asked all of us to discuss the question: Is it through nature or nurture that humans develop an attraction or aversion to certain forms of landscape? Taking as an example an aversion to heights (such as to ledges, cliffs, and precipices), some students argued the nature side: that biology had to be at work, for if fear of such landscapes are not established by childhood, the chances of surviving and thereby propagating one's genes would diminish. Other students argued the nurture side: that adults all the time protect infants from and warn children about the dangers of such landscapes. In such ways, it was a good discussion. But—even though we had just studied constructivist educational theory—not once did anyone suggest that their choice between nature or nurture was incomplete, and could perhaps be complemented by a constructivist account. No one, for example, suggested that infants and young children experiment every day with locomotion, with navigating through the physical world. They take tumbles, sometimes hard ones, and construct understandings of distance and spatial relationships, and of how some physical forms (like steps) can advance their purposes and others (like precipices) are to be avoided and feared. Development of such a constructivist account (compare Adolph 1997) requires that we break out of the dichotomy between nature and nurture.

During another part of this seminar I asked students to conduct some pilot research. After all, if we take seriously the constructivist dictum that to understand is to invent, then for students to understand the research enterprise it is not enough for them to read the work of other researchers; they need to do research themselves. Yet here, too, exogenous assumptions easily came into play. For example, one research group—who viewed themselves as working within a constructivist tradition—conducted a twenty-question bioregional survey with high school students. Questions included: Name three types of birds that live in your neighborhood. Name three types of trees that grow in your neighborhood. Name the nearest mountain. Where does your water come from? Where does your garbage go? This research group found that as a whole high school students could answer very few of the bioregional questions, although this effect was more pronounced for urban than rural and suburban students. They also reported that they shared the results of one classroom survey with that classroom's science teacher, and he was profusely ashamed of his students' performance, almost to the point of tears. The research group concurred with the teacher's disappointment. Based on their research, this group concluded (cautiously, given their small sample size) that more bioregional environmental education was needed.

As I wrote in chapter 5, survey methods can serve certain purposes very well; but they do not provide a powerful means toward understanding the depth, complexity, and coordinations in the development of knowledge. I would also suggest here that such surveys often lead educators astray. Why? Because surveys focus all too easily on content. Let us assume, for example, that we are that distressed science teacher. Or that we are the principal of the school (or the public at large), similarly distressed by the low scores of a large number of students on the bioregional "test." Teachers then often feel pressured to teach for the test. If students do not know the names of three birds, let us teach them by telling them, and by drill and practice. And if our education is successful our students will be able to perform better on the posttest. You can see, of course, that we are right back to traditional education: to instruction, not construction; to reinforcement, not interest. I will return shortly to a constructivist approach to bioregional education, and specific constructivist activities. My point here is that well-meaning educators and parents easily fall back on exogenous principles of learning, letting content knowledge—what DeVries (1997) calls "school varnish"—take the place of meaningful education.

Another difficulty some people have—especially those who are beginning to understand structural-developmental theory—is that they take hold of the idea of stages of knowledge and then too rigidly shape curriculum around them. For example, Piaget says abstract reasoning as characterized by formal operations does not usually begin until early adolescence. Beginning constructivist teachers sometimes interpret this finding to mean that they cannot engage in abstract reasoning with their preadolescent students. But when Piaget defines abstract reasoning in terms of formal operations he means something very specific: among other things, the ability to follow the logical train of an argument even when disagreeing with the premises, the ability to consider all possible outcomes, the ability to hold variables constant, and—in terms of structure—the consolidation of identity, negation, reciprocity, and correlation operations. But as I discussed at the end of chapter 8, younger children often engage in abstract reasoning, but of an earlier form. Accordingly, teachers should employ developmentally appropriate abstractions while teaching.

Implementing Constructivist Environmental Education

There is a chasm in the educational field. On one side, there are teachers who struggle daily with the demands of their vocation, such as developing lesson plans, correcting homework, reading student papers, and managing large numbers of students, let alone actually teaching, all the while being responsive to administrative constraints and parental concerns. On the other side, there are educational researchers and theorists who write about such issues. But often they use difficult language that is inaccessible to teachers, and have little actual classroom experience by which to translate theory into practice. Thus, toward helping bridge these two sides, I offer some ideas for how constructivist environmental education can actually take form in specific teaching contexts.

Consider first a set of activities—modified from Caduto and Bruchac (1988)—that a former student of mine, Ashley Weld, implemented successfully when walking "cold" into dozens of first- and second-grade classrooms (Kahn and Weld 1996). The activities follow from reading aloud Shel Silverstein's (1964) *The Giving Tree.* As the reader may know, this story is about a tree and boy who are friends. When the boy is little, he climbs up her trunk, and swings from her branches, and eats her apples. Together they play hide-and-go-seek. As he grows up, the boy begins to request material assistance from the tree. He asks her for her apples to sell for some money. She says she is happy to give. He asks for all of her branches to make a home for him and his family. She says she is happy to give. When he wants to travel to a distant land, he asks her for her trunk to build a canoe. Again, the tree gives. At the end of the story, the boy—now a man—comes back old and tired. He needs something, but the tree replies she has nothing left to offer him except a stump. The old man replies that that is all he needs now, and sits down.

It is a sad story. Yet it highlights, albeit in somewhat anthropomorphic terms, the reciprocity and intimacy that is possible with nature; and that while nature gives and gives, it can give too much, and perish, if we keep asking. And we, as humans, have kept asking.

In using this story in the early grades, it can work well to read it outside, under a tree. Then generate student discussion by asking, for example, What made you happy or sad? If you were the tree, what would you have liked to have happened? If you were the boy, what would you have liked to have happened? What did you like or not like about the ending? Ask the children why they think trees are important, and help bring out a diversity of ideas: Trees provide shade and food to animals and humans. Trees are fun to climb. When cut, trees provide lumber for human housing. When left standing, trees provide homes for animals. Perhaps sort children's answers into two categories: those that directly benefit humans and those that benefit a larger biotic community. Ask children to pretend to sit in front of a tree for a whole day, and ask them what things might they see, hear, and smell. Ask children to act out some of the occurrences: A bird feeding under a tree or flying to the top. A squirrel hopping from branch to branch. An old man taking rest in the shade. A little girl climbing the tree and picking an apple. Introduce children to how to make bark rubbings with crayons on the trunk of a tree, and allow time for their own creations. The same for leaf prints. Perhaps blindfold some of the children and have them feel the roots, trunk, branches, and leaves of different types of trees. Ask them to describe what they feel. How do trees differ? How are they the same?

Such stand alone activities can occur easily within a traditional educational context. But constructivist educators can move off school grounds, as well. For instance, consider again the topic of bioregional education. Where does our water come from? I know of one junior high school class in California that took this question seriously, and they spent a month following—by van and at times on foot—the trail of water (and politics) from San Francisco to the city's water source at the Hetch Hetchy reservoir. On the graduate level, Thomashow (1995) describes how, after about a month inside the classroom, he takes his students on a hike to the top of a New England mountain. From that perspective—having walked through "abandoned farmlands, stone walls, early successional growth, a few big white oaks and white pines . . . and finally a blueberry and boulder-covered summit" (p. 14)—he discusses bioregionalism, and later motivates student writing based on such direct experience.

Kohak (1984) gently captures the spirit of such education in his philosophical treatise on the moral sense of nature:

> Though philosophy must do much else as well, it must, initially, see and, thereafter, ground its speculation ever anew in seeing. So I have sought to see clearly and to articulate faithfully the moral sense of nature and of being human therein through the seasons lived in the solitude of the forest, beyond the powerline and the paved road, where the dusk comes softly and there still is night, pure between the glowing embers and the distant stars. . . . In writing of those years, I have not sought to "prove a point" but to evoke and to share a vision. Thus my primary tool has been the metaphor, not the argument, and the product of my labors is not a doctrine but an invitation to look and to see. (pp. xii–xiii)

Similarly, though environmental education must do much else as well, it must invite students to look and to see, not so as to acquire another "fact" about nature but rather to value it, through experiences lived and intimacy felt.

Such experiential education is practiced well by the National Audubon Expedition Institute. Students join a group of fifteen to twenty others and travel together as a community on a bus for a semester or more. The group typically concentrates in one or two areas of the United States, such as the Pacific Northwest, the deserts of New Mexico and Arizona, or the Maine coastal bioregion. Through their travels students and faculty may meet with environmental leaders, loggers, tree huggers, politicians, teachers, poets, urban poor, tribal leaders, environmental historians, fishermen, and more. They climb mountains, canoe rivers, and hike the desert. Most nights, they camp out. They feel the earth. See the stars. They make their group decisions by consensus, and through long meetings and heated discussions discover a fuller sense of community and a stronger sense of self. Through such grounded experience, they also work out analytically their position on environmental education and environmental ethics. Taken as a whole, they develop a deeper sense of what Thomashow (1995) calls an "ecological identity": "how people perceive themselves in reference to nature, as living and breathing beings connected to the rhythms of the earth, the biogeochemical cycles, the grand and complex diversity of ecological systems" (p. xiii).

If we believe that direct experience with nature is educationally essential—and I do—then we have a problem: what happens to nature when one million students from an urban area similarly investigate their own bioregion? Or worse, what would happen to nature were we to implement successfully such experiential education nationwide, let alone worldwide? Clearly wildness and nature as it currently exists would exist no longer. It is a troubling problem, and doubly troubling that a good educational approach may be self-defeating. Indeed, by seeking to prevent rampant use of nature, those (often active environmentalists) who currently enjoy natural settings can be charged with being elitist: seeking to preserve nature for only their enjoyment.

What can be done? First, it is essential that we not lose sight of what nature offers even within the urban context. Recall some of the qualitative data from the Houston parent study:

My grandson picks up all kinds of little animal things and some of them, I don't even know what they are myself, but he brings them in and gets a jar.

My kindergarten daughter, she might see something that looks injured and, um, she saw a worm. She doesn't pick up these black ones or brown ones because they sting. So this one was a yellow one and she said he was hungry. So she picked him up and took him over to a leaf and put him on it. You know, they do those type things.

Some of these Houston children were fascinated with the animals and vegetation within their reach: butterflies, ants, trees, worms, spiders, leaves, and flowers. Such results suggest that nature with distinction can be found everywhere, and that urban educators can look not only far off but close at hand for experiences from which to develop curriculum.

In chapter 6, where I report on the Houston child study, I noted that along with our psychological research we had established an educational partnership with an economically impoverished inner-city school. Through it, we sought in part to help teachers build on some of our psychological findings. Recall, for example, that we found that students understood about pollution in general, but had difficulty understanding that there was pollution in their own city. In response, the school's science teacher (Allan Smith) developed one activity where students coated strips of white paper with petroleum jelly. Then students would place these sticky strips of paper in various places around the school grounds. Some days later they would collect the strips, analyze the soot and grime that was on the paper, and compare differences between locations. Thus this science teacher sought ways to make this mostly "invisible" pollution real, in a concrete sense. He did something similar with water pollution. As noted in the Houston parent study, many parents were concerned with the quality of the water coming from their taps; some parents went so far as to buy bottled water, even given their very limited budget. With their parent's cooperation, the science teacher had older children bring in samples of their tap water, which they then tested for contaminants by conducting simple chemical experiments. Another informal finding from our research data was that many students thought that when you put garbage in a garbage can, the garbage disappeared forever. Thus teachers planned a field trip to a landfill. Some teachers also followed an idea of the principal (George Mundine): to have students make sculptures out of some material garbage brought from home—shifting from the idea of recycling to reusing.

Activities aside, there is larger issue here: How can environmental educational consultants offer meaningful help to urban schools? I suggest seven principles that were part of our Houston partnership. First, to whatever extent and in whatever forms possible, convey constructivist theory. Such conveying can take place through formal workshops or during curriculum development, or more informally in the hallways and lunchroom. Second, have the teachers themselves construct much of the curriculum. Third, build on the strengths, interests, reasoning, and values of the teachers, administration, and wider community. Fourth, build on the strengths, interests, reasoning, and values of the students. Fifth, if possible, establish means by which the school can garner money and recognition. As I noted in chapter 6, in the Houston project we wrote the grant so that half the moneys would go to the school to buy materials and to support release time and summer salaries for specific teachers and the principal. The principal also spoke about this collaboration at state-wide conferences. Sixth, have the school own the program when the consultation ends. I have seen many programs work relatively well while outsiders are involved; but once they leave, there is no lasting change because the program never became the school's own. Seventh, be content with small changes that are real. Small amounts of lasting change are usually better than large impacts that completely disappear soon after the consultation.

It is not enough, however, to provide urban children with good environmental education. We must also provide urban children with a good environment. Our cities need to be designed with nature in mind, in view, and within grasp. Kellert (1997) argues convincingly for this position. For, as follows from the diverse research that bears on the biophilia hypothesis, we need daily contact with nature not only for our physical but psychological well-being. Moreover, as Kellert (1998) argues, such contact needs to be enhanced by our buildings and physical infrastructure. My own collaborative research supports these positions insofar as children, adolescents, and adults across cultures spoke of the importance in their lives of animals, plants, trees, parks, and open spaces. Recall, for example, one of the students from Portugal who said: "[Gardens are important] because the city is a place that causes great stress and it gives a chance to someone to go to a place that is near, and to be in contact with nature, to stay calm." Or one of the black parents from Houston who said: "And it's a section of Alabama [Street] that I thought was so beautiful because of the trees, and they've cut down all the trees. And you know it hurts me every time I walk that way, and I hadn't realized that my son had paid attention to it, too." Thus environmental education lies in the hands not just of parents and educators, but of architects, builders, city planners, and politicians.

I had asked whether it is necessary for millions of urban youth to invade the wilds for authentic nature experiences. I have answered no, it is not. But I want to be careful here. For authentic experiences with nature are not coextensive with experiences in the wild. And wildness is sorely lacking in nature today, and increasingly lost—lost at a cost to the human psyche that too few people recognize. Turner (1996) writes: "I am concerned with preserving the authority of wild nature, or, more precisely, the authority of its presence in our experience and, hence, in the structure of our lives" (p. xiii). Shepard (1996) writes that "otherness of all kinds is in danger," and that such "otherness is essential to the discovery of the true self" (p. 5). Dean (1997) writes that "an enveloping wild landscape . . . [is] central to our original understanding of the world and our rightful place within it" (p. 17). As we continue to destroy wildness for the sake of material gain we might adapt to its loss, as we have already, no doubt. But it is a Faustian bargain, and we fool ourselves in believing otherwise.

Conclusion

Proponents of experiential environmental education sometimes dismiss the importance of intellectual inquiry. John Muir (1976) writes, for example:

I have a low opinion of books; they are but piles of stones set up to show coming travelers where other minds have been, or at best signal smokes to call attention. . . . No amount of word-making will ever make a single soul to know these mountains. As well seek to warm the naked and frost-bitten by lectures on caloric and pictures of flame. One day's exposure to mountains is better than cartloads of books. (p. 318)

But throughout this book I have argued something a little different: In fostering the human relationship with nature we need to pay attention not only to nature but to human nature—and it is deeply within our nature to use our intellects to construct increasingly sophisticated ideas, and to depend on them, physically and psychologically.

To elaborate on this point, I should like to say a few words about an episode from Primo Levi's ([1958] 1993) autobiographical account, *Surviving in Auschwitz*. In 1944 Levi was deported to Auschwitz. That was toward the end of the war, when the German government sought to make use of the labor of strong-bodied prisoners before eliminating them. Late in his account, Levi writes of a time when he and another prisoner were assigned to carry a hundred pound pot a half mile through the camp. The pot was attached to the middle of a pole, and the two prisoners each carried on their shoulders one side of the pole, thus half the weight. On this occasion, Levi discovered that he was paired with a kindred spirit, an educated person versed in the classics. There they are, trudging through the camp, almost collapsing under the weight of their burden, close to death, and Levi is divining meaning from Dante's poetry, and through its recitation:

Here, listen Pikolo, open your ears and your mind, you have to understand, for my sake:
"Think of your breed; for brutish ignorance
Your mettle was not made; you were made men,
To follow after knowledge and excellence."
As if I also was hearing it for the first time: like the blast of a trumpet, like the voice of God. For a moment I forget who I am and where I am. (p. 113)

While reciting another passage, Levi's memory comes up short. He can recall the beginning and ending, but try as he will he cannot remember the middle lines. Levi then turns Maslow's hierarchy on end: "I would give today's soup to know how to connect 'the like on any day' to the last lines" (p. 114).

That is the conception of the intellectual life we can offer our students, our children. The intellect's vitality. Its sheer beauty. Its power to heal and to sustain, and to help us create meaning. In our relationship with nature, let us not drive a wedge between the intellect and experience. Rather, by embracing both—which structural-developmental theory does—let us affirm what it means to be human in a world, if we choose wisely, of human goodness and natural splendor.

Appendix A: Interview Questions for the Portugal Study

(with Orlando Lourenço, University of Lisbon)

This appendix includes the complete set of structured questions that we asked (in Portuguese) of every participant in this study. As is standard practice when conducting semistructured interviews (see chap. 5), when appropriate we also followed each question with various probes. We would ask for clarification, offer possible counterexamples, and pose potential contradictions.

Prologue

1. In your opinion, are domestic animals [pets] important or not important? Why?

2. In your opinion, are wild animals important or not important? Why?

3. What's the difference in your relationship to domestic animals and wild animals?

4. In your opinion, are plants important or not important? Why?

5. In your opinion, are the gardens that exist around town important or not important? Why?

6. Do you know of or have you heard about any problems that affect the environment? Which ones?

7. Do you talk about the problems of the environment with your friends or with your family? If yes, what things come up in those conversations? If not, why don't you?

8. Do you do anything to protect the environment or to help solve some of its problems? If yes, what?

Part I: Water—The Case of the Polluted Waterway (the Rio Tejo)

9. Let's suppose that here in Lisbon a person threw their garbage into the Rio Tejo. Is that all right or not all right?

10. Let's say that in Lisbon everyone throws their garbage in the river; would that be all right or not all right?

11. Let's say that in Brazil everyone who lives near the Rio Amazônas throws their garbage into the Rio Amazônas, because that's one of the ways of solving the problem concerning where to dispose of trash. Is that all right or not all right?

12. Do you think throwing garbage in the Rio Tejo would affect the fish? How? Is that effect good, bad, both, or none of the above? Does it matter to you that fish would be affected this way? Why?

13. Do you think throwing garbage in the Rio Tejo would affect the birds? How? Is that effect good, bad, both, or none of the above? Does it matter to you that birds would be affected this way? Why?

14. Do you think throwing garbage in the Rio Tejo would affect the water? How? Is that effect good, bad, both, or none of the above? Does it matter to you that water would be affected this way? Why?

15. Do you think throwing garbage in the Rio Tejo would affect the view of the landscape? How? Is that effect good, bad, both, or none of the above? Does it matter to you that the view of the landscape would be affected this way? Why?

16. Do you think throwing garbage in the Rio Tejo would affect the people who live close to the river? How? Is that effect good, bad, both, or none of the above? Does it matter to you that people would be affected this way? Why?

Part II: Air—The Case of the Polluted Air

17. As you know, there is air pollution in Lisbon. Do you think this is a problem?

18. Do you think that driving a car increases the air pollution? How?

19. Do you think it is all right or not all right that a person drives his or her own car to work every day? Why? [If Yes: But how is it all right to drive the car if, as you said before, that increases the air pollution? If No: But how could this person arrive at his/her place of work? Would that be practical. Possible Probes: What if this person lives outside the city? What if there is no public transportation?]

20. Do you think it would be better if people didn't drive their cars to work? Why?

21. Let's suppose that the majority of people in Lisbon drive their cars to work. Is that all right or not all right? Why?

22. Do you think it would be better if nobody drove his/her car to work in Lisbon? Why?

23. Let's suppose that in New York City in the United States the majority of people drive their cars to work. Is that all right or not all right for people in New York City? Why?

24. Do you think it would be better if nobody drove his/her car to work in New York City? Why?

25. In Lisbon, do you think that there should be a law that would regulate pollution? If yes: Why? What would this law say? If not: Why not?

26. If you were the "ruler" of the world, what would you do to solve this problem of air pollution? Do you think that it would work? What else would you do?

Part III: Fire

27. As you know, last summer there were several huge forest fires in the forests of our country. Do you think that the fires in the forests were natural?

28. What does it mean to say that something is natural?

29. If a fire in the forest is caused by lightning, can we say that it is natural?

30. Do you think that a fire in the forest is natural if it is caused accidentally by a person?

31. Do you think that a fire in the forest is natural if it is set on purpose by a person?

32. If people want to live in harmony with nature, what should they do about fires in the forest?

Part IV: Earth

33. As you know, in several regions of our country trees are being cut down in the forests. Do you think this cutting causes any problem? Why?

34. In your opinion is it all right or not all right that people cut the forests? Why?

35. One classmate of yours with whom I talked said that cutting down the trees in the forest is all right because people need wood to build houses, and to make paper and other things that come from the trees. What do you think about what your classmate said? Why?

36. Another classmate to whom I spoke before told me that this cutting down of trees is not all right because it causes soil erosion. That is, the roots from trees hold the dirt and soil in place around them; after they are cut down and when it rains, the rain washes the top soil away. What do you think about what your classmate said? Why?

37. Are there any (other) problems caused by cutting down the trees in the forests? What?

38. If you were the "ruler" of the world, what would you do about the cutting down of the trees in the forests? Do you think that would work? What else would you do?

39. Is it natural for people to cut down the trees in the forests? Why?

40. Is it possible to live in harmony with nature and to cut down the trees in the forests? How?

41. For you, what does it mean to live in harmony with nature?

42. How do you know if someone is living in harmony with nature? Can you give an example?

Appendix B: Coding Manual for the Justification Data from the Portugal Study (Abridged Version)

(with Orlando Lourenço, University of Lisbon)

The justifications present a degree of difficulty in formalizing a coding system that does not arise with either the evaluative or content data. Thus some readers—particularly those who may want to pursue related lines of inquiry—will be interested in this appendix, which features the justification section from one of our coding manuals: The Portugal study (chap. 10).

In the categories described below, note that on occasion a category is followed with few or even no examples. In such cases, the reason is that few or no actual examples emerged from the half of the data set used to generate this manual. Nonetheless, the categories have been retained to be sensitive to such possible forms of reasoning in the second half of the data set and in future studies.

General Notes (adapted from Davidson, Turiel, and Black 1983):

1. Order of Analysis: Code each interview from beginning to end. This is because sometimes the participant or interviewer makes reference to reasoning generated at an earlier point in the interview, which the coder will want to recognize while coding.

2. Hesitancy of Response: Some participants are more prone than others to phrase the response in a hesitant manner—for example, "probably so," "maybe yes," and "perhaps not." Where the positive or negative connotation of such a response is clear, code as if it were expressed as a definite yes or no.

3. Misunderstanding: Do not code justifications that the participant uses on the basis of a misunderstanding of the question.

4. Countersuggestions: Sometimes a participant agrees with a counter-suggestion that is presented. But unless the participant restates or re-phrases this suggestion, or in some other way demonstrates his or her own understanding of it, do not use the participant's passive agreement as a criterion for coding.

5. Extent of Pertinent Information: In interpreting a response, do not hesitate—whenever needed and appropriate—to go beyond the particular question. Look at the responses to preceding and following questions, or consider the overall "tone" or quality of the interview. This does not mean doing a time-consuming search of the whole interview to solve numerous questions of detail; it does mean bearing in mind what the participant has said so far, and also being willing to go back and recode an earlier response if the participant says something later in the interview that helps clarify his or her earlier meaning.

6. Multiple Justifications: Either spontaneously or through probing, it is not uncommon for a participant to provide more than one justification for an evaluation. In such cases, code all significant justifications instead of only the most predominant one. Only two exceptions apply: (a) Should a participant repeatedly justify an evaluation using the same category, only code that category once for each evaluation; (b) If in justifying an evaluation there is both an elaborated and unelaborated justification, code only the elaborated justification.

7. Extent of Generic Categories:

Uncodable [O]: This category should be used in the following situations:

(1) The participant does not respond at all to the question.

(2) The response is too unintelligible or incomplete to code.

(3) Forcing the response into an existing category would substantially distort the meaning of the response.

(4) The participant answers a question other than the one asked. If such a response can be coded under the appropriate question elsewhere in the interview, do so.

(5) The justification to be coded follows an uncodable evaluation.

(6) The participant explicitly says that he does not know the answer and, in addition, has not provided an answer in the course of the interview.

Not Asked: This category should be used when the interviewer did not ask the question and no answer was, by chance, given to the question in the course of the interview.

1 Harm to Nature

A concern for the harm caused to nature. No reference is made to whether that concern derives from an anthropocentric or biocentric orientation. Notes: (a) This category is a minimal code, meaning that if it is used in conjunction with either anthropocentric or biocentric reasoning, then code only the latter reasoning. For example, the statement "It's wrong if a person throws trash in the Rio Tejo because it would pollute the water and it would look ugly" should just be coded as anthropocentric aesthetics (2.6), not harm to nonliving parts of nature (1.3).

1.1 Harm to Animals
An appeal is made to the welfare of animals. Notes: (a) If it is unclear whether the response refers to plants or animals (e.g., a reference to "living things") then code response as "harm to animals"—a more conservative code.

[It's not all right if a person throws trash in the Rio Tejo] *because . . . it would harm the fish.*

Why is it bad? The fish vanished . . . I think they all died.

[Wild animals are important because] *if there weren't some wild animals, others wouldn't have what* [they need] *to eat.*

1.2 Harm to Vegetation
An appeal is made to the welfare of plants, including such vegetation as grass, vegetables, flowers, and trees.

1.3 Harm to Nonliving Parts of Nature
An appeal is made to the welfare of nonliving parts of nature, including rocks, mountains, and oceans. While such parts may contain living things, the parts themselves are not living.

[It's not all right if a person throws trash in the Rio Tejo] *because it will pollute the water of the Tejo River.*

[It's not all right if a person throws trash in the Rio Tejo] *because it would harm the beaches.*

I think that it is wrong [if one person throws their trash in the Rio Tejo because] it is like helping to pollute the river. And not only the river, it is also the ground.

1.4 Harm to Species
An appeal is made that extends beyond a concern for the welfare of individual animals, or groups of animals, to a concern for the entirety of the group on the species level. Notes: (a) Do not double code this category with 1.5 (harm to natural processes), but code only 1.5.

Why is it bad? Because it causes the extinction of a flora species, another species of our planet.

Because their species would become extinct.

1.5 Harm to Natural Processes
An appeal is made that incorporates aspects of harm to animals, plants, nonliving parts of nature, and species, but embeds these aspects within an understanding of a larger systems perspective. Notes: (a) If reasoning of this form includes a reference to humans (e.g., "the oil would hurt plants, fish, bears, and humans") then code in its appropriate anthropocentric category.

1.5.1 Food Chain An appeal to the interconnected parts of the ecological community, largely by means of a descriptive hierarchy.

1.5.2 Ecosystem An appeal to a system of interconnectedness and dependencies that establish natural balances, sometimes though a discussion of habitat. Notes: (a) If a discussion of species includes reference to a systems perspective, code here.

[Deforestation is a problem] *because a person, when he cuts trees down causes that we have fewer plants, and plants generate oxygen, and that way there is less oxygen in the world.*

2 Anthropocentric

An appeal to how impacting the environment affects human beings. In other words, the environment is given consideration, but only because of human consequences. Notes: (a) Also code as anthropocentric if the justification refers to the avoidance of harm to people, e.g., "it's OK for bugs to die because then these bugs can't bite people."

2.1 Punishment Avoidance

An appeal to punishment or its avoidance. Notes (a) If punishment avoidance is embedded in dialogue with another justification (e.g., others' welfare), then code the other justification only. On the other hand, if punishment avoidance is juxtaposed with another justification, then code both.

2.2 Personal

An appeal to the personal predilections, interests, and projects of self and others. Notes: (a) Code only the highest order subcategory, e.g., if both 2.2.2 and 2.2.3 reasoning is invoked, code only 2.2.3.

2.2.1 Predilections
An appeal to largely unelaborated likings. Notes: (a) Minimal code; that is, do not use this code if any other anthropocentric or biocentric code is used.

[Domestic animals are important because] *concerning the human aspect I think they are nice.*

[It would matter to me if fish were harmed] *because I love fish.*

2.2.2 Interests
An appeal to fun, enjoyment, satisfaction, or recreation.

[Gardens are important] *for people to enjoy them.*

[Because] *we can send the children on Sundays, where they can throw bread crumbs to the fish. It is a lot of fun, they like it—I like it, too.*

[It's not all right for a person to throw water in the Rio Tejo] *because if the Rio Tejo were clean, we could swim in it.*

[Gardens are important] *because it is a good place to spend the afternoon.*

I would care very much [if the landscape was affected negatively] *because then it wouldn't be pleasant to go for a ride.*

2.2.3 Projects An appeal to interests that form part of a full-bodied conception of self.

[Gardens are important because] *people get to know each other in the gardens.*

[Gardens are important] *at least to hang out with friends. . . .*

[Gardens are important] *because it is a place where people can be together, walk, play, talk, and it is a place where the elderly use to go to entertain themselves.*

Gardens are very important to life because . . . without the gardens young people would choose to do other things, to spend their time in a different way. . . . For instance, a book that I've just finished reading is Os Filhos da Droga (Children of Drugs), *and one of the reasons that the young woman who wrote the book pointed out and showed throughout the book was the lack of places for young people to hang out to do their activities . . . I think we miss that, it is essential.*

2.3 Relational
An appeal to a relationship between humans and nature, but wherein the relationship fundamentally serves human needs.

2.3.1 Physical An appeal to one or more activities that structure a relationship.

[Domestic animals are important because] *we can play with them.*

2.3.2 Companionship An appeal to the benefits of companionship that nature accords to humans. Notes: (a) When physical and companionship reasons are combined, code only as companionship.

[Domestic animals are important] *because many of them become companions to those who own them, they are their friends, it is a help that is there to take care of them, it is company.*

[Domestic animals are important] *because they are companions to people.*

[Plants] *are important because as with the animals they keep us company.*

I like to have a relationship with domestic animals because, for instance, a domestic dog, I think that I can play with it, I always have a companion when, sometimes, I am by myself.

I consider a tree a companion when I am by myself, I can think about the good times I had with it.

2.3.3 Caretaking   An appeal to taking care of aspects of nature as one might take care of a person.

[Domestic animals are important because] *I love to take care of them.*

[Domestic animals are important] *because we can give love to animals.*

2.4   Welfare
An appeal to the welfare of human beings. Notes: (a) If physical welfare justifications overlap with material welfare, code only as physical.

2.4.1   Individual's Welfare   An appeal that is centrally concerned with the welfare of the self. Notes: (a) Only code as individual's welfare if it is clear that the appeal is self-centered, otherwise consider other's welfare.

2.4.1.1   Physical   An appeal to the self's physical welfare.

It is bad [to hurt the birds by means of pollution] *because I cause harm to the environment that is around me, I am causing harm to myself.*

From a personal point of view . . . I go to the beaches in the Coast, imagine that the seagull got some kind of disease, very contagious, and I caught it. So, what would I do later? Who is the culprit, I or the seagull? It would be me, wouldn't it? If it was I who started [the process of the] *disease, do you understand?*

[It's not all right that throwing trash in the Rio Tejo harms the fishermen because] *maybe the harm that happens to them might indirectly affect me.*

2.4.1.2   Material   An appeal to the self's material welfare.

2.4.1.3   Psychological   An appeal to the self's psychological welfare, including emotional states. Notes: (a) If the appeal has more of a relational sense to it, consider the companionship code or the biocentric relational code of psychological rapport, 3.4.1.

[Gardens are important because] *I speak for myself: When I am tired or feeling stressed, it is relaxing to sit under a tree, in a garden in the middle of town. . . .*

2.4.1.4 Educational An appeal to the self's educational welfare.

2.4.2 Others' Welfare An appeal to the welfare of other human beings. While the self can be included in this group, especially with inclusive pronouns (e.g., "we," "our," and "us"), the self is not the primary consideration.

2.4.2.1 Physical An appeal to the physical welfare of other human beings. Notes: (a) If the justification overlaps with material welfare, code only physical welfare.

Because it harms our health, generates bacteria, diseases, that people may catch.

Because the forests . . . give us oxygen, they help us a lot. If we destroy them, we are cutting our own oxygen and a possible way of protection they give us, and that is very bad for us.

[Plants are important] *because they generate oxygen. Without them we wouldn't live.*

[It would matter to me if the water was harmed] *because it would harm the health of everybody using that water either to drink or to bathe, anything at all.*

[Because] *green spaces give us clean air and that could disappear. For instance, there in Parque Monsanto, they are the lungs of the city. If we cut [the trees] down, we wouldn't have that anymore.*

2.4.2.2 Material An appeal to the material and economic welfare of human beings.

I would [care if the water were affected because] *look, again, it is a very selfish theory . . . From an economic point of view, the water would be captured and sent to a central plant where it would be treated. Who is paying for the process to clean the water? Isn't it us? So, we are causing harm to ourselves.*

[Plants are important because] *some give glue, give us paper, many things that we use in our daily life.*

[Domestic animals are important because] *in the case of the dog, they can guard a house when a thief comes, and such.*

Because . . . it throws gases [in the air] *. . . and there are important monuments along the Rio Tejo here in Portugal: Centro Cultural de Belem, Torre do Belem, Padrao dos Descobrimentos, and all those things.*

[It would matter to me if the view was affected] *because if that happens, it devalues many houses. People probably paid more in other times, and now they don't have money because of that.*

2.4.2.3 Psychological An appeal to the psychological welfare of human beings, including higher order emotional states, such as comfort, peace, security, calmness, and mental health.

[Gardens are important] *because the city is a place that causes great stress and it gives a chance to someone to go to a place that is near, and to be in contact with nature, to stay calm.*

[The problem with cutting down trees is] *that for me to go into a forest is something that brings peace and it is very healthy, for our mental health. I think we would lose a lot with that.*

[Gardens are important] *because I find a place to relax, a place to lower the stress level. . . .*

[Plants are important] *because they give us spiritual peace.*

[Gardens] *are important because in the middle of so much pollution and so many cars and so much stress, they are a way for people to relax. Usually when people spend some time in the garden, they feel, when they leave they feel more relaxed. It is important because they relieve stress.*

2.4.2.4 Educational An appeal to the potential for humans to learn from nature, and for nature to benefit others' overall development. Notes: (a) If the educational nature of the response is embedded within a concern for physical harm—e.g., the curing of a disease— code only physical welfare.

[Gardens are important] *mainly for the children, because they don't know, they've never seen a pig . . . and because of that at least they know what a tree is. It is important for that.*

[Gardens are important because] *children learn where things come from for real, because some children think that chickens come from the supermarket and things like that.*

Domestic animals are important mainly when there are children present because they promote their development.

[Gardens are important] *because that way we still have an idea of what life was some years ago, many years ago.*

2.4.3 Societal Welfare An appeal to the welfare of a society, nation, or planet as a whole.

Because . . . [certain trees] *are a national treasure, they are things that can't be seen anymore.*

2.4.4 Systemic Welfare An appeal that embeds concern for human welfare within a larger systemic context, including social systems (such as those that are political and economic), ecological systems (that appeal to how the welfare of humans is affected due to understanding how the interconnectedness and dependencies of ecological systems can establish natural balances across geography and time), and the coordination of social and economic systems. Notes: (a) This code does not pertain to individual biological systems such as the human body.

Those who own those little boats, and who use oil and throw the oil into the Tejo, that affects not only them, but . . . the fish are part of their system, and that influences their way of life. . . . They can't fish for two or three months, they start asking to pay later in the grocery store, their children have no clothes, do badly at school, their games are kind of weak because they lack energy. Those are concrete cases that lead to situations from a social point of view so low it is a very poor situation. All the population is affected.

Because we are causing air pollution and it is bad because of climate variations, because pollution meddles with the whole system, with those increases, those sudden variations in temperatures, all this has to do with air pollution. We, here in Lisbon, it has to do with us, it harms us directly.

2.4.5 Generational Welfare An appeal to the welfare of future generations. Notes: (a) This code overrides any particular type of welfare. For example, material generational welfare would be coded here, not under material welfare; (b) This category does not refer to past generations.

[Gardens are important] *knowing that the next generations are developing, being extremely important.*

When someone does that [throws garbage in the river] *it is polluting what is mine also, that it is also yours, that belongs to the coming generations, and not even saying that it is going to pollute the ocean that belongs to everybody.*

[Because] *it belongs to everybody, it is ours as a nation, it is ours as Europe, it is ours now, it is ours as* [it was] *our forefathers' and the next generation's.*

[I would worry if the fishermen were affected] *because sometimes I think that in our grandparents' time there were no such problems, and what now in our times is like that, in our children and grandchildren's time, how would it be? The world is going to end, it can't stand.*

2.5 Justice

An appeal that humans have rights, deserve respect, fair treatment, ownership of property, and/or merit freedom.

2.5.1 General A largely unelaborated appeal to rights, respect, fairness, ownership, and freedoms. Notes: (a) If societal considerations are part of justice, do not code as societal but only as justice.

[It's not all right if everyone in Lisbon threw trash in the Rio Tejo] *because it is polluting the water . . . and nobody has the right to make it dirty, it belongs to the public. Nobody, nobody, not even a group, not even by oneself. . . .*

We don't have the right to do that, it is going to destroy the ecosystem and we are part of the ecosystem. We are going to be harmed also, as it is happening already to us.

[It's wrong for a person to throw trash in the Rio Tejo] *because . . . the river doesn't belong to that person, and that person can do anything he wants in his own property. In the property that belongs to everybody, that belongs to the nation, that person, although he shares the ownership of the river, that river doesn't belong just to him, and that person has to think like that ad campaign that happened some years ago on TV—"Portugal Doesn't Belong Just to You." We can't do whatever comes to our minds.*

It is wrong [for a person to throw trash in the Rio Tejo] *because one has no right to make dirty what belongs to everybody.*

2.5.2 Justice Contextualized by Welfare An appeal that establishes fundamental linkages between justice and welfare reasoning, including appeals to unjustified harm.

[It would matter to me if the people were harmed by water pollution] *because I think that everybody in Portugal has the right to a minimum* [condition] *of living, and . . . this is a sign that they don't have* [those] *conditions to live.*

[It would matter to me if people were affected by throwing garbage in the river] *because people should be free, right? That is* [something] *necessary to everybody, and one's freedom means that they have* [certain] *conditions to live their lives.*

2.5.3 Isomorphic An appeal to a correspondence between two or more humans.

2.5.3.1 Direct Humans are viewed as essentially similar to one another, and sometimes the relevant properties are specified.

It is bad [if throwing trash in Rio Tejo harmed the fishermen] *because those people have the same right as we do to live in a pleasant environment. The same way that is harmful to us, it is harmful to them. . . . The same way that I want something good for me, I also want it to those people.*

[It would matter to me if the people along the Rio Tejo were harmed by the trash because] *they are living beings like me, and I wouldn't like it in my case.*

Yes, [I would worry if the fishermen were affected because] *I think that those people have the same rights that we do. We have the same rights, and I think that they should not be harmed.*

2.5.3.2 Conditional An appeal that establishes an if-then conditional judgment that leads to personal perspective-taking, though the judgment is not yet universalized.

2.5.4 Transmorphic An anthropocentric isomorphism is established, and then extended.

2.5.4.1 Compensatory Humans are viewed as both similar and different to one another, yet the qualities that establish similarity are overriding in establishing an anthropocentric justice orientation.

2.5.4.2 Hypothetical An appeal that establishes a hypothetical situation that leads through reflective moral impartiality to what is viewed to be a universally compelling conclusion.

[It would matter to me if people along the Rio Tejo were affected] *because they are people as I am. I think about those people as if I had to live there and I think that I would not feel very well.*

2.6 Aesthetic
An appeal to the preservation of the environment for the viewing or, more broadly, sensorial pleasure of humans.

[Plants are important to me] *because they are beautiful.*

[It would matter to me if the landscape was harmed by water pollution] *because . . . it is very unpleasant to be close to the river and* [feel] *that sensation.*

[The people by the river would be affected because] *the smell of the water, it should bother people to open their windows and feel that foul smell. . . .* [It would matter to me] *because a person doesn't have to smell dead fish or trash bags full of rotten stuff when she opens the window in the morning.*

[Plants are important because] *they give a better look to our environment, more color.*

[I would care if the fish were harmed because] *everybody likes to see fish swimming, pretty, those orange ones very pretty, and see them swimming.*

[It would matter to me if the water was harmed] *because . . . dirty water is unpleasant, there is no comparison to see a river with clean water, to see the fish swimming, to see the pebbles, and to see that brown, grayish, thick, disgusting water.*

3 Biocentric

An appeal to the moral standing of a larger ecological community of which humans may be a part.

3.1 Intrinsic Value of Nature
An appeal that nature has value, and the validity of that value is not derived solely from human interests. Notes: (a) The categories are

hierarchically integrated, so only code the highest category, with this one exception: categories 3.1.2 and 3.1.3 can be double-coded.

3.1.1 General An appeal to general, implicit, or unspecified values about nature ("Because the water is good"). Notes: (a) If unelaborated rights and respect reasoning is embedded in reasoning about values, then code as rights. (b) Minimal code.

3.1.2 Biological Life An appeal to the intrinsic value of biological life.

[Wild animals are important because] *every living being has to have the opportunity to be alive.*

[Plants are important] *because they are living beings, and I don't think that one should get close to a plant, and without any reason, step on it, crush it, kill it. . . . It is like they have life, it is not like a stone that is a dead being.*

[Wild animals are important because] *they are the essence, the basics.*

It is important to have some [wild animals] *in order to continue to have [them] always.*

3.1.3 Inanimate Objects An appeal to the intrinsic value of non-biological natural objects (e.g., water, rocks, and mountains), including appeals that derive their biocentrism through attributing feelings and emotions to natural objects (e.g., "the mountain doesn't want the cutters to cut down trees because it'd probably be, like, felt").

3.1.4 Natural Processes An appeal that is based on the validity and inherent worth of natural processes.

3.1.4.1 Food Chain An appeal to the intrinsic value of natural processes by means of a descriptively characterized food chain.

[Wild animals are important] *because they are part of our planet and have their place in the food chain.*

3.1.4.2 Ecosystem An appeal to a system of interconnectedness and dependencies that establish natural balances across geography and time.

[Wild animals are important] *because they represent a natural balance.*

[Wild animals are important] *in order to have a smooth operation of the ecosystem.*

[Wild animals are important because] *they are part of the living world. . . . It has to do with the structure that we are trying to destroy.*

[Wild animals] *are important because they maintain the balance of the ecosystem.*

[It's not all right to cut the trees] *because it destroys all the ecosystem that is interconnected. For instance, the animals that live around those trees, if they are cut down, they disappear and affect the whole ecosystem.*

[Wild animals are important] *because they keep the ecosystem working.*

3.1.5 Telos of Nature   An appeal to the design or purpose in natural objects, processes, or occurrences, including (a) appeals based on a sense of metaphysics, balance, completeness, aesthetics, or religion, and (b) appeals to the original state of nature (an "is") to justify its continuation (an "ought").

[Wild animals are important] *because . . . for sure they have a definite role. We don't know for sure, but I think that they have their own reason to exist.*

I think that [wild animals] *are important because if someone created them it is because they have some kind of role, therefore they are important.*

[It is bad to throw trash in the Rio Tejo] *because the water, the ocean was not created to have trash thrown into it.*

[Wild animals] *are important because they are a patrimony of humanity. They are the result of millions of years of evolution. Of course they are important.*

Because the river was not created [for people] *to throw trash into it, it is a natural means, another natural means that should not be destroyed.*

3.1.6 Isomorphic An appeal that is based on recognizing a correspondence between the intrinsic value of humans and those of other natural biological or nonbiological entities.

3.1.6.1 Direct Humans and nature are viewed as essentially similar, and sometimes the relevant properties are specified. Accordingly, an appeal is made that nature thereby deserves the same values considerations as humans.

[Domestic animals are important] *because they are living beings, like people.*

Every living being has its own importance, haven't they? We can't say that we are superior to all the others. They are as important as we are. . . .

[Plants] *are important, as the animals are important, because they are living beings, and live like us.*

I would care about the rest of the species as much as I would about human beings. About the birds, the fish, and everything else. But man is selfish, he thinks more about himself, but I would worry about that. Would you worry more about man? *No, I wouldn't, it would be the same. I think that yes, we are living beings, fish are another living being, birds are another living being.*

3.1.6.2 Conditional An appeal that establishes an if-then conditional judgment of value that leads to personal perspective-taking, though the judgment is not yet universalized.

[Because] *if we don't like to live surrounded by trash,* [the fish] *don't like it also.*

Because people did that [to the fish] *and, for instance, people do that, but if they were fish, they wouldn't like that* [someone] *did that to them.*

3.1.7 Transmorphic A biocentric isomorphism is established, and then extended. Included here are reasons wherein an inequivalent correspondence between nature and humans does not void the mapping of similar value considerations to nature.

3.1.7.1 Compensatory Humans and nature are viewed as both similar and different, yet the qualities that establish similarity are overriding in establishing a biocentric justice orientation.

[Wild animals] *are very important because we are not the ones who rule this planet. . . . I think that only because we have a mind and we are able to think, we don't have, we are not better than the other animals, domestic or wild.*

[Wild animals are important] *because they breathe like we do, and sometimes we think that because they are animals, they are not like us, that they don't do certain things. Then we end up seeing that they do.*

3.1.7.2 Hypothetical An appeal that establishes a hypothetical situation which leads through reflective moral impartiality to what is viewed to be a universally compelling conclusion.

3.2 Harmony
An appeal to a biocentric conception of harmony between humans and nature.

3.2.1 Relational An appeal to a harmonious relationship between humans and nature.

3.2.1.1 Psychological Rapport An appeal to an authentic intimacy between the person and the natural.

3.2.1.2 Preservation An appeal to a less personal, more generalized form of caretaking for nature.

[I would worry if fish were affected] *because when we have an aquarium at home, we like the fish to . . . be well cared for. That is why I would like the rest of the fish to be well cared for.*

[Wild animals are important] *because I think that we have to preserve something in this world.*

3.2.2 Compositional An appeal to being in balance with nature, through a sense of either proportion (including proportion, moderation, and magnitude of harm), or equality or equity (through direct or indirect compensation or through an abstract understanding of fairness, i.e., not necessarily "tit for tat").

[Because] *it is not going to be in harmony, it is going, there will be a lack of balance.*

3.3 Justice
An appeal that nature has rights, deserves respect or fair treatment, or merits freedom.

3.3.1 General   An unelaborated appeal to rights and respect.

It is bad because as I said before, every living being has its own rights, and the fish have the right of being in the sea, and no human being has the right of taking away their habitat.

[Wild animals are important because] *they also have their rights . . . right to live.*

[Animals] *are also part of the world, they have a right to live.*

It is wrong [to cut the trees of a forest down] *because it disrespects the life of a plant.*

Because . . . birds also have the right to be birds going around, flying and singing, even if they are seagulls.

[Pets are important] *because they also have the right to be free.*

[Wild animals are important] *because I think that all animals have the right to live their life.*

3.3.2 Justice Contextualized by Welfare   An appeal that establishes fundamental linkages between justice and welfare reasoning, including an appeal to a right to freedom from unjustified harm, or to the innocence of an agent that has moral standing. Notes: (a) This category can be double-coded with either isomorphic (3.3.3) or transmorphic (3.3.4) justifications.

I am worried [about fish being hurt by pollution, because] *the fish are living in a polluted environment and what right have I to, well, what right have I as a human being to influence the life of another living being when it, poor thing, hasn't caused me any harm?*

[It's not all right to cut the trees because] *it causes the destruction of the habitat, it causes the poor trees* [to die] *that have no fault whatsoever. . . .*

[It would matter to me if birds were harmed] *because I think they have a right to live; I think that they haven't done any harm, so I should not do harm to them either.*

3.3.3 Isomorphic An appeal that is based on recognizing a justice correspondence between humans and other natural biological or non-biological entities.

3.3.3.1 Direct Humans and nature are viewed as essentially similar, and sometimes the relevant properties are specified. Accordingly, an appeal is made that nature thereby deserves the same rights and respect as humans.

I think that the fish have a right to live, to a good life. We are not more not less than the animals that were here before us. They have a right to have a good life.

[It would matter to me that birds were harmed by pollution because] *they are animals that have as much right to live as I do.*

[It's not all right that the community in Lisbon threw garbage in the Rio Tejo because] *it would destroy the environment, and we don't have the right to do that, because we are living beings the same as the others.*

[It would matter to me if the fish were harmed because] *I think that the fish have the right to live as the rest of us* [do].

Because I think that [wild animals] *also have the right to live in the jungle . . . It is not just us that have to live . . . Because I think that in the same way that we procreate, they also have the right to live, to be happy. . . . Because I think that they were also created the same way that we were, and because we have the right to live, everybody— I think everybody has a right to live.*

[It would matter to me if birds were harmed] *because I think that the animals have as much right to live and to have good conditions of life as we do, and the pollution that affects us will affect them also.*

3.3.3.2 Conditional An appeal that establishes an if-then conditional judgment that leads to personal perspective-taking, though the judgment is not yet universalized.

3.3.4 Transmorphic A biocentric isomorphism is established, and then extended. Included here are reasons wherein an inequivalent correspondence between nature and humans does not void the mapping of similar biocentric justice considerations.

3.3.4.1 Compensatory Humans and nature are viewed as both similar and different, yet the qualities that establish similarity are overriding in establishing a biocentric justice orientation.

3.3.4.2 Hypothetical An appeal that establishes a hypothetical situation which leads through reflective moral impartiality to what is viewed to be a universally compelling conclusion.

References

A conversation with Richard Ofshe (April 1994). *California Monthly,* 104(5): 20–24.

Abram, D. 1996. *The spell of the sensuous: Perception and language in a more-than-human world.* New York: Pantheon.

Adolph, K. E. 1997. Learning in the development of infant locomotion. *Monographs of the Society for Research in Child Development,* 62(3, Serial No. 251).

Aristotle 1962. *Nichomachean ethics.* Translated by M. Oswald. Indianapolis, Indiana: Bobbs-Merrill.

Armstrong, J. 1996. Lost in space. *Orion,* 15(2): 22–27.

Arsenio, W. F. 1988. Children's conceptions of the situational affective consequences of sociomoral events. *Child Development,* 59: 1611–1622.

Arsenio, W., and Lover, A. 1995. Children's conceptions of sociomoral affect: Happy victimizers, mixed emotions, and other expectancies. In M. Killen and D. Hart (Eds.), *Morality in everyday life: Developmental perspectives* (pp. 87–128). New York: Cambridge University Press.

Asch, S. E. 1952. *Social psychology.* Englewood Cliffs, NJ: Prentice-Hall.

Baldwin, J. M. [1897] 1973. *Social and ethical interpretations in mental development.* Reprint, New York: Arno.

Balling, J. D., and Falk, J. H. 1982. Development of visual preference for natural environments. *Environment and Behavior,* 14: 5–38.

Bandura, A. 1977. *Social learning theory.* Englewood Cliffs, NJ: Prentice-Hall.

Bandura, A., and Walters, R. 1963. *Social learning theory and personality development.* New York: MacMillan.

Barash, D. 1979. *The whisperings within.* London: Penguin.

Barenboim, C. 1975. Developmental changes in the interpersonal cognitive system from middle childhood to adolescence. *Child Development,* 48: 1467–1474.

Bass, E., and Davis, L. 1988. *The courage to heal.* New York: Harper Collins.

Baxter, W. F. 1986. People or penguins. In D. VanDeVeer and C. Pierce (Eds.), *People, penguins, and plastic trees* (pp. 214–218). Belmont, CA: Wadsworth.

Beck, A., and Katcher, A. 1996. *Between pets and people.* West Lafayette, IN: Purdue University Press.

Bennett, W. J., and Delatree, E. J. 1978. Moral education in the schools. *The Public Interest,* 50: 81–98.

Beresky, A.E. (Ed.) 1991. *Fodor's Brazil.* New York: Fodor's Travel.

Beringer, A. 1994. *The moral ideals of care and respect: A hermeneutic inquiry into adolescents' environmental ethics and moral functioning.* Frankfurt: Peter Lang Publishing.

Berkowitz, M., Guerra, N., and Nucci, L. 1991. Sociomoral development and drug and alcohol abuse. In W. Kurtines and J. Gewirtz (Eds.), *Handbook of moral behavior and development,* Vol. 3. (pp. 35–53). Hillsdale, NJ: Lawrence Erlbaum.

Berry, W. 1977. *The unsettling of America: Culture and agriculture.* New York: Avon.

Biaggio, A. M. B. 1994. Moral development and attitudes toward ecology. Paper presented at the International Association for Cross Cultural Psychology Symposium on Eco-ethical Thinking in Cultural Perspective. Universitat des Saarlandes, Germany. July–August.

Biaggio, A. M. B. 1997. Kohlberg and the "just community": Developing ethical sense and citizenship in the schools. *Psicologia Reflexão e Crítica,* 10: 47–70.

Bjorklund, D. F., and Kipp, K. 1996. Parental investment theory and gender differences in the evolution of inhibition mechanisms. *Psychological Bulletin,* 120: 163–188.

Blasi, A. 1980. Bridging moral cognition and moral action: A critical review of the literature. *Psychological Bulletin,* 1: 1–45.

Blight, J. G. 1981. Must psychoanalysis retreat to hermeneutics?: Psychoanalytic theory in the light of Popper's evolutionary epistemology. *Psychoanalysis and Contemporary Thought,* 4: 147–206.

Boyd, D. R. 1989. The character of moral development. In L. Nucci (Ed.), *Moral development and character education: A dialogue* (pp. 95–123). Berkeley: McCutchan.

Bryant, B., and Mohai, P. (Eds.). 1992. *Race and the incidence of environmental hazards: A time for discourse.* Boulder, CO: Westview Press.

Bullard, R. D. 1987. *Invisible Houston: The black experience in boom and bust.* College Station, TX: Texas A and M University Press.

Bullard, R. D. 1990. *Dumping in Dixie: Race, class, and environmental quality.* Boulder, CO: Westview Press.

Bunting, T. E., and Cousins, L. R. 1985. Environmental dispositions among school-age children. *Environment and Behavior,* 17(6): 725–768.

Buss, D. M. 1992. Mate preference mechanisms: Consequences for partner choice and intrasexual competition. In J. H. Barkow, L. Cosmides, and J. Tooby (Eds.),

The adapted mind: Evolutionary psychology and the generation of culture (pp. 249–266). New York: Oxford University Press.

Caduto, M. 1983. Toward a comprehensive strategy for environmental values education. *Journal of Environmental Education*, 14(4): 12–18.

Caduto, M., and Bruchac, J. 1988. *Keepers of the earth: Native American stories and environmental activities for children*. Golden, CO: Fulcrum.

Callicott, J. B. 1985. Intrinsic value, quantum theory, and environmental ethics. *Environmental Ethics* 7: 257–275.

Campbell, R. L., and Bickhard, M. H. 1987. A deconstruction of Fodor's anticonstructivism. *Human Development*, 30: 48–59.

Campbell, R. L., and Bickhard, M. H. 1992. Clearing the ground: Foundational questions once again. *Journal of Pragmatics*, 17: 557–602.

Campbell, R. L., and Christopher, J. C. 1996. Moral development theory: A critique of its Kantian presuppositions. *Developmental Review*, 16: 1–47.

Case, R. 1992. *The Mind's Staircase*. Hillsdale, NJ: Lawrence Erlbaum.

Chandler, M. 1997. Stumping for progress in a post-modern world. In E. Amsel and K. Ann Renninger (Eds.), *Change and development: Issues of theory, methods, and application* (pp. 1–26). Mahwah, NJ: Lawrence Erlbaum.

Chapman, M. 1988. *Constructive evolution: Origins and development of Piaget's thought*. Cambridge: Cambridge University Press.

Chawla, L. 1988. Children's concern for the natural environment. *Children's Environments Quarterly*, 5(3): 13–20.

Childers, P., and Wimmer, M. 1971. The concept of death in early childhood. *Child Development*, 42: 1299–1301.

Chomsky, N. 1959. A review of B. F. Skinner's *Verbal Behavior. Language*, 35: 26–58.

Cleaver, E. 1969. *Post-prison writings and speeches*. Edited by R. Scheer. New York: Random House.

Colby, A., and Damon, W. 1992. *Some do care: Contemporary lives of moral commitment*. New York: Free Press.

Coley, J. D. 1995. Emerging differentiation of folkbiology and folkpsychology: Attributions of biological and psychological properties to living things. *Child Development*, 66: 1856–1874.

Coley, J. D., Medin, D. L., and Atran, S. 1997. Does rank have its privilege?: Inductive inferences within folkbiological taxonomies. *Cognition*, 64: 73–112.

Cosmides, L., Tooby, J., and Barkow, J. H. 1992. Introduction: Evolutionary psychology and conceptual integration. In J. H. Barkow, L. Cosmides, and J. Tooby (Eds.), *The adapted mind: Evolutionary psychology and the generation of culture* (pp. 3–15). New York: Oxford University Press.

Coss, R. G. 1990. Picture perception and patient stress: A study of anxiety reduction and postoperative stability. Unpublished manuscript, Department of Psychology, University of California, Davis.

Cowan, P. A. 1978. *Piaget with feeling.* New York: Holt, Rinehart and Winston.

Crenson, M. A. 1971. *The un-politics of air pollutions.* Baltimore: Johns Hopkins University Press.

Crews, F. 1986. In the big house of theory. *The New York Review of Books,* May 29, pp. 36–42.

Crews, F. 1989. The parting of the Twains. *The New York Review of Books,* June 20, pp. 39–44.

Crews, F. 1995. *The memory wars: Freud's legacy in dispute.* New York: New York Review of Books.

Culler, J. 1982. *On deconstruction: Theory and criticism after structuralism.* Ithaca, NY: Cornell University Press.

Daily, G., and Ehrlich, P. 1997–98. Population extinction and the biodiversity crisis. *Wild Earth,* 7(4): 35–45.

Damon, W. 1977. *The social world of the child.* San Francisco: Jossey-Bass.

Davidson, P., Turiel, E., and Black, A. 1983. The effect of stimulus familiarity on the use of criteria and justifications in children's social reasoning. *British Journal of Developmental Psychology,* 1: 49–65.

Dawkins, R. 1976. *The selfish gene.* New York: Oxford University Press.

Dean, B. (1997). A multitude of witnesses. *Northern Lights,* 13(1): 14–17.

Derrida, J. (1978). *Writing and difference.* London: Routledge and Kegan Paul.

Derrida, J. 1993. l'Affaire Derrida. Letter to the editor. *The New York Review of Books,* February 11, p. 44.

DeVries, R. 1987. Moral development in play. In D. Bergen (Ed.), *Play as a learning context* (pp. 103–105). Exeter, NH: Heinemann.

DeVries, R. 1988. *Constructivist education.* Paper presented to the Association for Constructivist Teaching, West Point Academy. October.

DeVries, R. (1997). Piaget's social theory. *Educational Researcher,* 26(2): pp. 4–17.

DeVries, R., and Kohlberg, L. 1990. *Constructivist early education: Overview and comparison with other programs.* Washington, DC: National Association for the Education of Young Children.

DeVries, R., and Zan, B. 1994. *Moral classrooms, moral children: Creating a constructivist atmosphere in early education.* New York: Teachers College.

Dewey, J. [1916] 1966. *Democracy and education.* Reprint, New York: Macmillan.

Diamond, J. 1993. New Guineans and their natural world. In S. R. Kellert and E. O. Wilson (Eds.), *The Biophilia hypothesis* (pp. 251–271). Washington, DC: Island Press.

Dunker, K. 1939. Ethical relativity?: An enquiry into the psychology of ethics. *Mind,* 48: 39–57.

Eisenberg, N. 1989. The development of prosocial values. In N. Eisenberg, J. Reykowski, and E. Staub (Eds.), *Social and moral values.* Hillsdale, NJ: Lawrence Erlbaum.

Eisenberg, N., and Fabes, R. 1998. Prosocial development. In N. Eisenberg (Ed.), *Social, Emotional, and Personality Development* (pp. 710–778). Vol 3 of W. Damon (Ed.), *Handbook of child psychology.* 5th ed. New York: Wiley.

Eisenberg-Berg, N. 1979. Development of children's prosocial moral judgment. *Developmental Psychology* 15: 128–137.

Ellis, B. J. 1992. The evolution of sexual attraction: Evaluative mechanisms in women. In J. H. Barkow, L. Cosmides, and J. Tooby (Eds.), *The adapted mind: Evolutionary psychology and the generation of culture* (pp. 267–288). New York: Oxford University Press.

Encarta® 98 Desk Encyclopedia © & sP 1996–97 Microsoft Corporation.

The First Issue. 1993. *The Wilderness Society,* 56(200): 6.

Fischer, C. S. 1994. Widespread likings: Review of the biophilia hypothesis. *Science,* 263: 1161–1162.

Fischer, V., Boyle, J., Schulman, M., and Bucuvalas, M. 1980. *A survey of the public's attitudes toward soil, water, and renewable resources conservation policy.* Washington, DC: U.S. Government Printing Office.

Fish, S. 1996. Professor Sokal's bad joke. *New York Times,* May 21, p. A23.

Fishkin, J. S. 1982. *The limits of obligation.* New Haven: Yale University Press.

Flavell, J. H. 1963. *The developmental psychology of Jean Piaget.* New York: Van Nostrand.

Foucault, M. 1980. Truth and power. In C. Gordon (Ed.), *Power/knowledge: selected interviews and other writings 1972–77* (pp. 109–133). New York: Pantheon.

Freud, S. [1923] 1960. *The ego and the id.* Trans. by J. Riviere. New York: Norton.

Freud, S. [1930] 1961. *Civilization and its discontents.* Trans. by J. Strachey. New York: Norton.

Freud, S. [1924] 1963. The passing of the Oedipus-complex. Trans. by J. Riviere. In P. Rieff (Ed.), *Sexuality and the Psychology of Love* (pp. 176–182). New York: Collier.

Freud, S. [1920] 1967. *Beyond the pleasure principle.* Trans. by J. Strachey. New York: Bantam.

Friedman, B. 1997. Social judgments and technological innovation: Adolescents' conceptions of property, privacy, and electronic information. *Computers in Human Behavior,* 13: 327–351.

Gadgil, M. 1993. Of life and artifacts. In S.R. and E. O. Wilson (Eds.), *The Biophilia hypothesis* (pp. 365–377). Washington, DC: Island Press.

Garrod, A., Beal, C., and Shin, P. 1990. The development of moral orientation in elementary school children. *Sex Roles* 22: 13–27.

Gates, H. L., Jr., and West, C. 1996. The gap between black leaders and their constituents. *The Chronicle of Higher Education.* April 5, p. B7.

Gaylord, C. E., and Bell, E. 1995. Environmental justice: A national priority. In L. Westra and P. S. Wenz (Eds.), *Faces of environmental racism* (pp. 29–39). Lanham, MD: Rowman and Littlefield.

Gewirth, A. 1978. *Reason and morality.* Chicago: University of Chicago Press.

Gilardi, J. 1994. Valdez oil spill recovery "remarkable," Exxon chief says. *The Journal of Commerce and Commercial Bulletin,* May 3, p. 7B.

Gilligan, C. 1982. *In a different voice.* Cambridge, MA: Harvard University Press.

Ginsburg, H. P. 1997. *Entering the child's mind: The clinical interview in psychological research and practice.* Cambridge: Cambridge University Press.

Ginsburg, H. P., and Opper, S. 1969. *Piaget's theory of intellectual development.* Englewood Cliffs, NJ: Prentice-Hall.

Giroux, H. A. 1988. *Schooling and the struggle for public life.* Minneapolis: University of Minnesota Press.

Giroux, H. A. 1990. *Curriculum discourse as postmodernist critical practice.* Victoria, Australia: Deakin University Press.

Glassman, M., and Zan, B. 1995. Moral activity and domain theory: An alternative interpretation of research with young children. *Developmental Review,* 1: 434–457.

Goodman, N. 1972. *Problems and projects.* New York: Bobbs-Merrill.

Gough, N. 1987. Learning with environments: Towards an ecological paradigm for education. In I. Robottom (Ed.), *Environmental education: Practice and possibility* (pp. 49–67). Victoria, Australia: Deakin University Press.

Grant, D. S., II 1997. Religion and the left: The prospects of a green coalition. *Environmental Values,* 19: 115–134.

Greenway, R. 1995. The wilderness effect and ecopsychology. In T. Roszak, M. E. Gomes, and A. D. Kanner (Eds.), *Ecopsychology: Restoring the earth, healing the mind* (pp. 122–135). San Francisco: Sierra Club Books.

Griffin, D. R. 1992. Introduction to SUNY series in constructive postmodern thought. In D. W. Orr, *Ecological literacy: Education and the transition to a postmodern world.* Albany: State University of New York Press.

Grünbaum, A. 1984. *The foundations of psychoanalysis.* Berkeley and Los Angeles: University of California Press.

Haidt, J., Koller, S. H., and Dias, M. G. 1993. Affect, culture, and morality, or is it wrong to eat your dog? *Journal of Personality and Social Psychology,* 65: 613–628.

Hamilton, W. D. 1964. The genetical evolution of social behavior. *The Journal of Theoretical Biology,* 7: 1–16.

Hammer, R., and MacLaren, P. 1991. Rethinking the dialectic: A social semiotic perspective for educators. *Educational Theory,* 41(1): 23–46.

Harris, J. 1992. *Against relativism: A philosophical defense*. LaSalle, IL: Open Court.

Hart, D., Yates, M., Fegley, S., and Wilson, G. 1995. Moral commitment in inner-city adolescents. In M. Killen and D. Hart (Eds.), *Morality in everyday life: Developmental perspectives* (pp. 317–341). New York: Cambridge University Press.

Hart, R. A. 1997. *Children's participation: The theory and practice of involving young citizens in community development and environmental care*. New York: UNICEF.

Hassan, I. 1985. The culture of postmodernism. *Theory, culture, and society, 2*: 119–131.

Hatch, E. 1983. *Culture and morality*. New York: Columbia University Press.

Hayles, N. K. 1995. Searching for common ground. In M. E. Soule and G. Lease (Eds.), *Reinventing nature?: Responses to postmodern deconstruction* (pp. 47–63). Washington, DC: Island Press.

Heerwagen, J. 1990. The psychological aspects of windows and window design. In K. H. Anthony, J. Choi, and B. Orland (Eds.), *Proceedings of the 21st annual conference of the Environmental Design Research Association*. Oklahoma City: EDRA.

Helwig, C. C. 1995. Adolescents' and young adults' conceptions of civil liberties: Freedom of speech and religion. *Child Development, 66*: 152–166.

Helwig, C. C. 1997. Making moral cognition respectable (again): A retrospective review of Lawrence Kohlberg. *Contemporary Psychology, 42*: 191–195.

Helwig, C. C., Tisak, M., and Turiel, E. 1990. Children's social reasoning in context. *Child Development, 61*: 2068–2078.

Hershey, M. R., and Hill, D. B. (1977–78). Is pollution "a White thing?": Racial differences in preadults' attitudes. *Public Opinion Quarterly, 41*: 439–458.

Hogan, R. 1975. Theoretical egocentrism and the problem of compliance. *American Psychologist, 30*: 533–539.

Hohm, C. F. 1976. A human-ecological approach to the reality and perception of air pollution: The Los Angeles Case. *Pacific Sociological Review, 19*: 21–44.

Hollos, M., Leis, P. E., and Turiel, E. 1986. Social reasoning in Ijo children and adolescents in Nigerian communities. *Journal of Cross-Cultural Psychology, 17*: 352–374.

Holloway, M. 1991. Soiled shores. *Scientific American*, October, 102–116.

hooks, b. 1996. Touching the earth. *Orion*, 15(4): 21–22.

Houston, P. 1996. The bear in the woods, the bear in us. *New York Times*, OP-ED, June 22.

Howe, D., Kahn, P. H., Jr., and Friedman, B. 1996. Along the Rio Negro: Brazilian children's environmental views and values. *Developmental Psychology, 32*: 979–987.

Hoy, D. C. 1985. Interpreting the law: Hermeneutical and poststructuralist perspectives. *Southern California Law Review,* 58: 135–176.

Huberty, C. J. 1987. On statistical testing. *Educational Researcher,* 16: 4–9.

Huebner, A., and Garrod, A. 1991. Moral reasoning in a karmic world. *Human Development,* 34: 341–352.

Hume, D. [1748] 1961. *An enquiry concerning human understanding.* Edited by L. Selby-Bigge. Reprint, Oxford: Clarendon.

Hume, D. [1751] 1983. *An enquiry concerning the principles of morals.* Edited by J. B. Schneewind. Reprint, Indianapolis, IN: Hackett Publishing Company.

Hungerford, H. R. 1975. Myths of environmental education. *Journal of Environmental Education,* 7(2): 21–26.

Hunt, L. H. 1987. Generosity and the diversity of the virtues. In R. B. Kruschwitz and R. C. Roberts (Eds.), *The virtues: Contemporary essays on moral character* (pp. 217–228). Belmont, CA: Wadsworth.

Irvine, S. 1997–98). The great denial: Puncturing pro-natalist myths. *Wild Earth,* 7(4): 8–17.

Jarrett, J. L. 1957. *The quest for beauty.* Englewood Cliffs, NJ: Prentice-Hall.

Jarrett, J. L. 1991. *The teaching of values: Caring and appreciation.* New York: Routledge.

Jinks, J. L. 1975. A total curricular approach to environmental education. *Journal of Environmental Education,* 7(2): 11–20.

Kahn, P. H., Jr. 1991. Bounding the controversies: Foundational issues in the study of moral development. *Human Development,* 34: 325–340.

Kahn, P. H., Jr. 1992. Children's obligatory and discretionary moral judgments. *Child Development,* 63: 416–430.

Kahn, P. H., Jr. 1994. Resolving environmental disputes: Litigation, mediation, and the courting of ethical community. *Environmental Values,* 3: 211–228.

Kahn, P. H., Jr. 1995. Commentary on D. Moshman's, "The construction of moral rationality." *Human Development,* 38: 282–288.

Kahn, P. H., Jr. 1997a. Developmental psychology and the biophilia hypothesis: Children's affiliation with nature. *Developmental Review,* 17: 1–61.

Kahn, P. H., Jr. 1997b. Children's moral and ecological reasoning about the Prince William Sound oil spill. *Developmental Psychology,* 33: 1091–1096.

Kahn, P. H., Jr. 1997c. Bayous and jungle rivers: Cross-cultural perspectives on children's environmental moral reasoning. In H. Saltzstein (Ed.), *Culture as a context for moral development: New perspectives on the particular and the universal* (pp. 23–36). New Directions for Child Development (W. Damon, Series Editor). San Francisco: Jossey-Bass.

Kahn, P. H., Jr., and Friedman, B. 1995. Environmental views and values of children in an inner-city Black community. *Child Development,* 66: 1403–1417.

Kahn, P. H., Jr., and Friedman, B. 1998. On nature and environmental education: Black parents speak from the inner city. *Environmental Education Research,* 4: 25–39.

Kahn, P. H., Jr., and Lourenço, O. 1999. *Air, water, fire, and earth: A developmental study in Portugal of environmental conceptions and values.* Journal article in preparation.

Kahn, P. H., Jr., and Lourenço, O. in press. *Reinstating modernity in social science research—or—The status of Bullwinkle in a post-postmodern era. Human Development.*

Kahn, P. H., Jr., and Turiel, E. 1988. Children's conceptions of trust in the context of social expectations. *Merrill-Palmer Quarterly,* 34: 403–419.

Kahn, P. H., Jr., and Weld, A. 1996. Environmental education: Toward an intimacy with nature. *Interdisciplinary Studies in Literature and Environment,* 3(2): 165–168.

Kant, I. [1785] 1964. *Groundwork of the metaphysic of morals.* Translated by H. J. Paton. Reprint, New York: Harper Torchbooks.

Kaplan, R. 1973. Some psychological benefits of gardening. *Environment and Behavior,* 5: 145–152.

Kaplan, R. 1977. Patterns of environmental preference. *Environment and Behavior,* 9: 195–216.

Kaplan, R. 1985. Nature at the doorstep: Residential satisfaction and the nearby environment. *Journal of Architectural and Planning Research,* 2: 115–127.

Kaplan, R., and Kaplan, S. 1989. *The experience of nature: A psychological perspective.* Cambridge: Cambridge University Press.

Kaplan, S. 1983. A model of person-environment compatibility. *Environment and Behavior,* 15: 311–332.

Kaplan, S. 1987. Aesthetics, affect, and cognition: Environmental preferences from an evolutionary perspective. *Environment and Behavior,* 19: 3–32.

Kaplan, S. 1992. Environmental preference in a knowledge-seeking, knowledge-using organism. In J. H. Barkow, L. Cosmides, and J. Tooby (Eds.), *The adapted mind: Evolutionary psychology and the generation of culture* (pp. 581–598). New York: Oxford University Press.

Kaplan, S. 1995. Review of S. R. Kellert and E. O. Wilson, *The biophilia hypothesis. Environment and Behavior,* 27: 801–804.

Katcher, A., Friedmann, E., Beck, A., and Lynch, J. 1983. Looking, talking, and blood pressure: The physiological consequences of interaction with the living environment. In A. Katcher and A. Beck (Eds.), *New perspectives on our lives with companion animals.* Philadelphia: University of Pennsylvania Press.

Katcher, A., Segal, H., and Beck, A. 1984. Comparison of contemplation and hypnosis for the reduction of anxiety and discomfort during dental surgery. *American Journal of Clinical Hypnosis,* 27: 14–21.

Katcher, A., and Wilkins, G. 1993. Dialogue with animals: Its nature and culture. In S. R. Kellert and E. O. Wilson (Eds.), *The biophilia hypothesis* (pp. 173–197). Washington, DC: Island Press.

Keeble. J. 1993. A parable of oil and water. *The Amicus Journal,* 15: 35–42.

Keller, M., and Edelstein, W. 1991. The development of socio-moral meaning making: Domains, categories, and perspective-taking. In W. Kurtines and J. Gewirtz (Eds.), *Handbook of moral behavior and development, Vol. 2* (pp. 89–114). Hillsdale, NJ: Lawrence Erlbaum.

Kellert, S. R. 1980. Contemporary values of wildlife in American society. In W. Shaw and I. Zube (Eds.), *Wildlife values* (pp. 31–37). Ft. Collins: U.S. Forest Service.

Kellert, S. R. 1985. Attitudes toward animals: Age related development among children. *Journal of Environmental Education,* 16: 29–39.

Kellert, S. R. 1991. Japanese perceptions of wildlife. *Conservation Biology,* 5: 297–308.

Kellert, S. R. 1993. The biological basis for human values of nature. In S. R. Kellert and E. O. Wilson (Eds.), *The biophilia hypothesis* (pp. 42–69). Washington, DC: Island Press.

Kellert, S. R. 1994. Unpublished letter to the editor of *Science.*

Kellert, S. R. 1995. Concepts of nature East and West. In M. E. Soule and G. Lease (Eds.), *Reinventing nature?: Responses to postmodern deconstruction* (pp. 103–121). Washington, DC: Island Press.

Kellert, S. R. 1996. *The value of life.* Washington, DC: Island Press.

Kellert, S. R. 1997. *Kinship to mastery: biophilia in human evolution and development.* Washington, DC: Island Press.

Kellert, S. R. 1998. Ecological challenge, human values of nature and sustainability in the built environment. Paper presented at the Rinker Eminent Lecture Series, University of Florida. January.

Kellert, S. R., and Wilson, E. O. (Eds.). 1993. *The biophilia hypothesis.* Washington, DC: Island Press.

Kempton, W., Boster, J. S., and Hartley, J. A. 1996. *Environmental values in American culture.* Cambridge, MA: MIT Press.

Killen, M. 1990. Children's evaluations of morality in the context of peer, teacher-child, and familial relations. *Journal of Genetic Psychology, 151:* 395–410.

Killen, M. 1996. Justice and care: Dichotomies or coexistence? *Journal for a Just and Caring Education,* 2(1): 42–58.

Killen, M., and Hart, D. (Eds.). 1995. *Morality in everyday life: Developmental perspectives.* New York: Cambridge University Press.

Killen, M., Leviton, M., and Cahill, J. 1991. Adolescent reasoning about drug use. *Journal of Adolescent Research,* 6: 336–356.

Kitcher, P. 1985. *Vaulting ambition: Sociobiology and the quest for human nature.* Cambridge, MA: MIT Press.

Kohak, E. 1984. *The embers and the stars: A philosophical inquiry into the moral sense of nature.* Chicago: University of Chicago Press.

Kohlberg, L. 1969. Stage and sequence: The cognitive-developmental approach to socialization. In D. A. Goslin (Ed.), *Handbook of socialization theory and research* (pp. 347–480). New York: Rand McNally.

Kohlberg, L. 1971. From is to ought: How to commit the naturalistic fallacy and get away with it in the study of moral development. In T. Mischel (Ed.), *Psychology and genetic epistemology* (pp. 151–235). New York: Academic Press.

Kohlberg, L. 1974. Education, moral development and faith. *Journal of Moral Education, 4:* 5–16.

Kohlberg, L. 1980. High school democracy and educating for a just society. In R. L. Mosher (Ed.), *Moral education: A first generation of research* (pp. 20–57). New York: Praeger.

Kohlberg, L. 1984. *Essays on moral development: The psychology of moral development.* Vol. 2. San Francisco: Harper and Row.

Kohlberg, L. 1985. The just community approach to moral education in theory and practice. In M. W. Berkowitz and F. Oser (Eds.), *Moral education: Theory and application* (pp. 27–87). Hillsdale, NJ: Lawrence Erlbaum.

Kohlberg, L. and Mayer, R. 1972. Development as the aim of education. *Harvard Educational Review, 42:* 449–496.

Kreger, J. 1973. Ecology and black student opinion. *Journal of Environmental Education, 4:* 30–34.

Kurtines, W. M., and Gewirtz, J. L. (Eds.). 1991. *Morality, moral behavior and moral development.* Vols. 1–3. New York: John Wiley.

Lakoff, G. 1987. *Women, fire, and dangerous things: What categories reveal about the mind.* Chicago: University of Chicago Press.

Langer, J. 1969. *Theories of development.* New York: Holt, Rinehart, and Winston.

Langer, S. K. [1937] 1953. *An introduction to symbolic logic.* Reprint, New York: Dover.

Lapsley, D. K. 1996. Commentary. *Human Development, 39:* 100–107.

Laupa, M. 1991. Children's reasoning about three authority attributes: Adult status, knowledge, and social position. *Developmental Psychology, 27:* 321–329.

Laupa, M., and Turiel, E. 1986. Children's conceptions of adult and peer authority. *Child Development, 57:* 405–412.

Lawrence, E. A. 1993. The sacred bee, the filthy pig, and the bat out of hell: Animal symbolism as cognitive biophilia. In S. R. Kellert and E. O. Wilson (Eds.), *The biophilia hypothesis* (pp. 301–341). Washington, DC: Island Press.

Leopold, A. [1949] 1970. *A Sand Country Almanac.* Reprint, New York: Ballantine Books.

Levi, P. [1958] 1993. *Survival in Auschwitz*. Reprint, New York: Macmillan.

Lewis, G. E. 1981–82. A review of classroom methodologies for environmental education. *Journal of Environmental Education*, 13(2): 12–15.

López, A., Atran, S., Coley, J. D., Medin, D. L., and Smith, E. E. 1997. The tree of life: Universals of folkbiological taxonomies and inductions. *Cognitive Psychology*, 32: 251–295.

Lourenço, O. 1990. From cost-perception to gain-construction: Toward a Piagetian explanation of the development of altruism in children. *International Journal of Behavioral Development*, 13: 119–132.

Lourenço, O. 1991. Is the care orientation distinct from the justice orientation?: Some empirical data in ten- to-eleven-year-old children. *Archives de Psychologi*, 59: 17–30.

Lourenço, O. 1996. Reflections on narrative approaches to moral development. *Human Development*, 39: 83–99.

Lourenço, O. and Machado, A. 1996. In defense of Piaget's theory: A reply to 10 common criticisms. *Psychological Review*, 103: 143–164.

Madden T. 1992. Cultural factors and assumptions in social reasoning in India. Unpublished doctoral dissertation, University of California, Berkeley.

Marascuilo, L. A., and McSweeney, M. 1977. *Nonparametric and distribution-free methods for the social sciences*. Monterey, CA: Brooks/Cole.

Marascuilo, L. A., Omelich, C. L., and Gokhale, D. V. 1988. Planned and post hoc methods for multiple sample McNemar tests with missing data. *Psychological Bulletin*, 103: 238–245.

Maslow, A. H. 1975. *Motivation and personality*. New York: Viking.

Masson, J. M. 1984. *The assault on truth: Freud's suppression of the seduction theory*. New York: Farrar, Straus and Giroux.

Matthews, D. 1993. The remarkable recovery of Prince William Sound. *The Lamp*, 4–13.

McGillicuddy-DeLisi, A., Watkins, C., and Vinchur, A. 1994. The effect of relationship on children's distributive justice reasoning. *Child Development*, 65: 1694–1700.

McKibben, B. 1998. *Maybe one: A personal and environmental argument for single-child families*. New York: Simon and Schuster.

McLaren, P. 1989. On ideology and education: Critical pedagogy and the cultural politics of resistance. In H. A. Giroux and P. McLaren (Eds.), *Critical pedagogy, the state, and cultural struggle* (pp. 174–202). Albany: State University of New York Press.

McMillen, L. 1993. "L'Affaire Derrida" pits theorist who founded deconstruction against editor of book on Heidegger's role in Nazi era. *The Chronicle of Higher Education*, February 17, p. A8.

Medin, D. L, and Atran, S. (Ed.) In press. *Folkbiology*. Cambridge, MA: MIT Press.

Medin, D. L., Lynch, E. B., Coley, J. D., and Atran, S. 1997. Categorization and reasoning among tree experts: Do all roads lead to Rome? *Cognitive Psychology,* 32: 49–96.

Mei, Y. P. 1972. Mo Tzu. In P. Edwards (Ed.), *The encyclopedia of philosophy,* Vol. 5. (pp. 409–410). New York: Macmillan.

Miller, A. 1981. Integrative thinking as a goal of environmental education. *Journal of Environmental Education,* 12(4): 3–8.

Miller, J. G. 1994. Cultural diversity in the morality of caring: Individually oriented versus duty-based interpersonal moral codes. *Cross-Cultural Research,* 28: 3–39.

Mills, S. 1997–98. Nulliparity and a cruel hoax revisited. *Wild Earth,* 7(4): 18–21.

Milton, J. [1674] 1978. *Paradise lost.* In M. Y. Hughes (Ed.), *John Milton: complete poems and major prose* (pp. 211–469). Indianapolis, IN: Odyssey Press.

Mitchell, R. C. 1979. Public opinion on environmental issues. In *Environmental quality* (pp. 401–425). Washington, DC: Council on Environmental Quality.

Mohai, P. 1990. Black environmentalism. *Social Science Quarterly,* 4: 744–765.

Mohai, P. 1997. Gender differences in the perception of most important environmental problems. *Race, Gender and Class,* 5: 153–169.

Mohai, P., and Bryant, B. 1992. Environmental racism: Reviewing the evidence. In B. Bryant and P. Mohai (Eds.), *Race and the incidence of environmental hazards: A time for discourse* (pp. 163–176). Boulder, CO: Westview Press.

Mohai, P., and Bryant, B. 1996. Is there a "race" effect on concern for environmental quality? Paper presented at the Meeting of the Rural Sociological Society, Des Moines, Iowa. August.

Mohai, P., and Twight, B. W. 1987. Age and environmentalism: An elaboration of the Buttel model using national survey evidence. *Social Science Quarterly,* 68: 798–815.

Moore, E. O. 1982. A prison environment's effect on health care service demands. *Journal of Environmental Systems,* 11: 17–34.

Moore, G. E. [1903] 1978. *Principia ethica.* Reprint, Cambridge: Cambridge University Press.

Morss, J. R. 1992. Making waves: Deconstruction and developmental psychology. *Theory and Psychology,* 2: 445–465.

Moshman, D. 1995. The construction of moral rationality. *Human Development* 38: 265–281.

Muir, J. (1976). The philosophy of John Muir. In E. W. Teale (Ed.), *The wilderness world of John Muir* (pp. 311–323). Boston: Houghton Mifflin.

Mumford, L. (1970). *The conduct of life.* New York: Harcourt Brace Jovanovich.

Murphy, J. W. (1988). Computerization, postmodern epistemology, and reading in the postmodern era. *Educational Theory,* 38: 175–182.

Myers, G. 1996. Review of S. R. Kellert and E. O. Wilson's, *The biophilia Hypothesis. Environmental Ethics* 18: 327–330.

Myers, G. 1998. *Children and animals: Social development and our connections to other species.* Boulder, CO: Westview Press.

Nabhan, G. P. 1995. Cultural parallax in viewing North American habitats. In M. E. Soule and G. Lease (Eds.). *Reinventing nature?: Responses to postmodern deconstruction* (pp. 87–101). Washington, DC: Island Press.

Nabhan, G. P., and St. Antoine, S. 1993. The loss of floral and faunal story: The extinction of experience. In S. R. Kellert and E. O. Wilson (Eds.), *The biophilia hypothesis* (pp. 229–250). Washington, DC: Island Press.

Nabhan, G. P., and Trimble, S. 1994. *The geography of childhood: Why children need wild places.* Boston: Beacon Press.

Neill, A. S. [1960] 1977. *Summerhill.* Reprint, New York: Simon and Schuster.

Nelson, R. K. 1983. *Make prayers to the raven: A Koyukon view of the northern forest.* Chicago: University of Chicago Press.

Nelson, R. 1989. *The island within.* New York: Random House.

Nelson, R. 1993. Searching for the lost arrow: Physical and spiritual ecology in the hunter's world. In S. R. Kellert and E. O. Wilson (Eds.), *The biophilia hypothesis* (pp. 201–228). Washington, DC: Island Press.

Nevers, P., Gebhard, U., and Billmann-Mahecha, E. 1997. Patterns of reasoning exhibited by children and adolescents in response to moral dilemmas involving plants, animals, and ecosystems. *Journal of Moral Education,* 26: 169–186.

Niederman, R. D. 1978. Development of the social group in childhood and adolescence. Unpublished doctoral dissertation, University of California, Berkeley.

Nisan, M. 1991. The moral balance model: Theory and research extending our understanding of moral choice and deviation. In W. Kurtines and J. Gewirtz (Eds.), *Handbook of moral behavior and development,* Vol. 3 (pp. 213–249). Hillsdale, NJ: Lawrence Erlbaum.

Noddings, N. 1984. *Caring: A feminine approach to ethics and moral education.* Berkeley: University of California Press.

Norris, C. 1982. *Deconstruction: Theory and practice.* New York: Methuen.

Nucci, L. P. 1981. The development of personal concepts: A domain distinct from moral and societal concepts. *Child Development,* 52: 114–121.

Nucci, L. P. 1985. Children's conceptions of morality, social conventions, and religious prescriptions. In C. G. Harding (Ed.), *Moral dilemmas: Philosophical and psychological reconsiderations in the development of moral reasoning* (pp. 137–174). Chicago: Precedent Press.

Nucci, L. 1996. Morality and the personal sphere of actions. In E. S. Reed, E. Turiel, and T. Brown (Eds.), *Values and knowledge* (pp. 41–60). Mahwah, NJ: Lawrence Erlbaum.

Nucci, L. (1997). Culture, universals, and the personal. In H. Saltzstein (Ed.), *Culture as a context for moral development: New perspectives on the particular and the universal* (pp. 5–22). New Directions for Child Development (W. Damon, Series Editor). San Francisco: Jossey-Bass.

Nyrop, R. F. (Ed.). 1983. *Brazil: A country study.* Washington, DC: American University Press.

Ogbu, J. U. 1977. Racial stratification and education: The case of Stockton, California. *IRCD Bulletin,* 12: 1–26.

Ogbu, J. U. 1990. Overcoming racial barriers to equal access. In J. I. Goodlad and P. Keating (Eds.), *Access to knowledge: An agenda for our nation's schools* (pp. 59–89). New York: College Entrance Examination Board.

Ogbu, J. U. 1993. Differences in cultural frame of reference. *International Journal of Behavioral Development,* 16(3): 483–506.

Orians, G. H., and Heerwagen, J. H. 1992. Evolved responses to landscapes. In J. H. Barkow, L. Cosmides, and J. Tooby (Eds.), *The adapted mind: Evolutionary psychology and the generation of culture* (555–579). New York: Oxford University Press.

Orr, D. W. 1992. *Ecological literacy: Education and the transition to a postmodern world.* Albany: State University of New York Press.

Orr, D. W. 1993. Love it or lose it: The coming biophilia revolution. In S. R. Kellert and E. O. Wilson (Eds.), *The biophilia hypothesis* (pp. 415–440). Washington, DC: Island Press.

Orr, D. W. 1994. *Earth in mind.* Washington, DC: Island Press.

Oser, F., and Althof, W. 1993. Trust in advance: On the professional morality of teachers. *Journal of Moral Education,* 22: 253–275.

Ostheimer, J. M., and Ritt, L. G. 1976. *Environment, energy, and black Americans.* Beverly Hills: Sage.

Pain, S. 1993. The two faces of the Exxon disaster. *New Scientist,* May, pp. 11–13.

Palinkas, L. A., Petterson, J. S., Russell, J., and Downs, M. A. 1993. Community patterns of psychiatric disorders after the Exxon Valdez oil spill. *American Journal of Psychiatry,* 150: 1517–1523.

Parental role in learning is urged. 1994. *Boston Globe,* September 8, p. 3.

Partridge, E. 1984. Nature as a moral resource. *Environmental Ethics,* 6: 101–130.

Partridge, E. 1996. Ecological morality and nonmoral sentiments. *Environmental Ethics,* 18: 149–163.

Peterson, D. J. 1993. *Troubled lands: The legacy of Soviet environmental destruction.* Boulder, CO: Westview.

Piaget, J. [1929] 1960. *The child's conception of the world.* Reprint, Totowa, NJ: Littlefield, Adams and Co.

Piaget, J. [1932] 1969. *The moral judgment of the child.* Reprint, Glencoe, IL: Free Press.

Piaget, J. [1952] 1963. *The origins of intelligence in children.* Reprint, New York: Norton.

Piaget, J. [1952] 1965. *The child's conception of number.* Reprint, New York: Norton.

Piaget, J. 1970. *Structuralism.* New York: Harper and Row.

Piaget, J. 1971a. *Insights and illusions of philosophy.* New York: World.

Piaget, J. 1971b. *Genetic epistemology.* New York: Norton.

Piaget, J. 1971c. *Psychology and epistemology.* New York: Viking.

Piaget, J. 1983. Piaget's theory. In W. Kessen (Ed.), *History, Theory, and Methods* (pp. 1030–128). Vol. 1 of *Handbook of child psychology,* edited by P. H. Mussen. 4th ed. New York: Wiley.

Pinker, S. 1997. *How the mind works.* New York: Norton.

Pinker, S., and Bloom, P. 1992. Natural language and natural selection. In J. H. Barkow, L. Cosmides, and J. Tooby (Eds.), *The adapted mind: Evolutionary psychology and the generation of culture* (pp. 451–493). New York: Oxford University Press.

Plato. 1956. *Meno.* In E. H. Warmington and P. G. Rouse (Eds.), *Great dialogues of Plato* (pp. 28–68). Trans. by W. H. D. Rouse. New York: Signet.

Pomerantz, G. 1986. Environmental education tools for elementary schoolchildren: The use of a popular children's magazine. *Journal of Environmental Education, 17*(4): 17–22.

Potter, E. 1989. *The best of Brazil.* New York: Crown.

Puka, B. (Ed.). (1995). *Moral development: A compendium.* 7 vols. Hamden, CT: Garland.

Rabb, G. 1993. Heini Hediger—A pioneer in the science of animal behavior. *Der Zoologische Garten,* 63: 163–167.

Radke-Yarrow, M., Zahn-Waxler, C., and Chapman, M. 1983. Children's prosocial dispositions and behavior. In P. H. Mussen (Ed.), *Handbook of child psychology.* New York: John Wiley.

Rawls, J. 1971. *A theory of justice.* Cambridge, MA: Harvard University Press.

Reed, E. S. 1996. *Encountering the world: Toward an ecological psychology.* New York: Oxford University Press.

Regan, T. 1983. *The case for animal rights.* Berkeley: University of California Press.

Regan, T. 1986. The case for animal rights. In D. VanDeVeer and C. Pierce (Eds.), *People, penguins, and plastic trees* (pp. 32–39). Belmont, CA: Wadsworth.

Reiss, B. 1990. Halt, por favor! Forest Police! *Outside,* July, pp. 30–36, 90–96.

Rest, J., and Narvaez, D. 1991. The college experience and moral development. In W. Kurtines and J. Gewirtz (Eds.), *Handbook of moral behavior and development*, Vol. 2 (pp. 229–245). Hillsdale, NJ: Lawrence Erlbaum.

Richardson, L. 1988. The collective story: Postmodernism and the writing of sociology. *Sociological Focus*, 21: 199–207.

Rogoff, B. 1990. *Apprenticeship in thinking: Cognitive development in social context*. New York: Oxford.

Robottom, I. M. (Ed.). 1987. *Environmental education: Practice and possibility.* Victoria, Australia: Deakin University Press.

Rohwer, W. D., Jr., Rohwer, C. P., and B-Howe, J. R. 1980. *Educational psychology: Teaching for student diversity.* New York: Holt, Rinehart, and Winston.

Rolston, H., III 1981. Values in nature. *Environmental Ethics,* 3: 113–128.

Rolston, H., III. 1989. *Philosophy gone wild.* Buffalo, NY: Prometheus Books.

Rolston, H., III. 1993. Biophilia, selfish genes, shared values. In S. R. Kellert and E. O. Wilson (Eds.), *The biophilia hypothesis* (pp. 381–414). Washington, DC: Island Press.

Rolston, H., III. 1994. Value in nature and the nature of value. In R. Attfield and A. Belsey (Eds.), *Philosophy and the natural environment* (pp. 13–30). Cambridge: Cambridge University Press.

Rolston, H., III. 1995. Does aesthetic appreciation of landscapes need to be science-based? *British Journal of Aesthetics,* 35: 374–386.

Rolston, H., III. 1997. Nature for real: Is nature a social construct? In T. D. J. Chappell (Ed.), *The philosophy of the environment* (pp. 38–64). Edinburgh: University of Edinburgh Press.

Rosenau, P. M. 1992. *Post-modernism and the social sciences: Insights, inroads, and intrusions.* Princeton, NJ: Princeton University Press.

Rosenthal, A. M. 1995. You are Palden Gyatso. *New York Times*, April 11, p. A25.

Roszak, T. 1993. *The voice of the earth: An exploration of ecopsychology.* New York: Touchstone.

Rothenberg, D. 1993. *Hand's end: Technology and the limits of nature.* Berkeley: University of California Press

Rousseau, J. J. 1964. *Emile, Julie, and other writings.* R. L. Archer (Ed.). Woodbury, N.Y.: Barron's Educational Series.

Rousseau, J. J. [1972] 1979. *Emile.* Translated by A. Bloom. Reprint, New York: Basic Books.

Ruse, M., and Wilson, E. O. 1985. The evolution of ethics. *New Scientist,* 108(17): 50–52.

Rushton, J. P. 1982. Social learning theory and the development of prosocial behavior. In N. Eisenberg (Ed.), *The development of prosocial behavior* (pp. 77–105). New York: Academic Press.

Russians struggle to clean up spill. 1994. *New York Times*, November 11, p. A14.

Ryan, K. 1989. In defense of character education. In L. Nucci (Ed.), *Moral development and character education: A dialogue* (pp. 3–17). Berkeley: McCutchan.

Sasson, J. 1992. *Princess: A true story of life behind the veil in Saudi Arabia*. New York: William Morrow.

Saxe, G. B. 1990. *Culture and cognitive development: Studies in mathematical understanding*. Hillsdale, NJ: Lawrence Erlbaum.

Scheffler, S. 1986. Morality's demands and their limits. *The Journal of Philosophy*, 10: 531–537.

Scheffler, S. 1992. *Human morality*. New York: Oxford University Press.

Scholes, R. 1989. *Protocols of reading*. New Haven: Yale University Press.

Seabrook, J. 1991. The David Lynch of architecture. *Vanity Fair*, January, pp. 74–79, 125–129.

Searle, J. R. 1983. The word turned upside down. *New York Review of Books*, October 27, pp. 74–79.

Searle, J. R. 1990. Is the brain's mind a computer program? *Scientific American*, 262: 26–31.

Selman, R. L. 1980. *The growth of interpersonal understanding*. New York: Academic Press.

Selman, R. L., Jaquette, D., and Lavin, D. R. 1977. Interpersonal awareness in children: toward an integration of developmental and clinical child psychology. *American Journal of Orthopsychiatry*, 47: 264–274.

Shepard, P. 1993. On animal friends. In S. R. Kellert and E. O. Wilson (Eds.), *The biophilia hypothesis* (pp. 275–300). Washington, DC: Island Press.

Shepard, P. 1995. Virtually hunting reality in the forests of simulacra. In M. E. Soule and G. Lease (Eds.), *Reinventing nature?: Responses to postmodern deconstruction* (pp. 17–29). Washington, DC: Island Press.

Shepard, P. 1996. *The others: How animals made us human*. Washington, DC: Island Press.

Shweder, R. A. 1986. Divergent rationalities. In D. W. Fiske and R. A. Shweder (Eds.), *Metatheory in social science: Pluralisms and subjectivities* (pp. 163–196). Chicago: University of Chicago Press.

Shweder, R. A., Mahapatra, M., and Miller, J. B. 1987. Culture and moral development. In J. Kagan and S. Lamb (Eds.), *The emergence of morality in young children* (pp. 1–82). Chicago: University of Chicago Press.

Silverstein, S. 1964. *The giving tree*. New York: Harper and Row.

Simon, H. 1969. The architecture of complexity. In H. Simon (Ed.), *The sciences of the artificial* (pp. 84–118). Cambridge, MA: MIT Press.

Skinner, B. F. 1971. *Beyond freedom and dignity.* New York: Alfred A. Knopf.

Skinner, B. F. 1974. *About behaviorism.* New York: Knopf.

Skinner, B. F. [1948] 1976. *Walden two.* Reprint, New York: Macmillan.

Smetana, J. G. 1982. *Concepts of self and morality: Women's reasoning about abortion.* New York: Praeger.

Smetana, J. G. 1983. Social-cognitive development: Domain distinctions and coordinations. *Developmental Review,* 3: 131–147.

Smetana, J. G. 1984. Does morality have a gender?: A commentary on Pratt, Golding, and Hunter. *Merrill-Palmer Quarterly,* 30: 341–348.

Smetana, J. G. 1995. Morality in context: Abstractions, ambiguities and applications. In R. Vasta (Ed.), *Annals of child development,* Vol. 10 (pp. 83–130). London: Jessica Kingsley.

Smetana, J. 1997. Parenting and the development of social knowledge reconceptualized: A social domain analysis. In J. E. Grusec and L. Kuczynski (Eds.), *Handbook of parenting and the transmission of values* (pp. 162–192). New York: Wiley.

Smetana, J. G., Killen, M., and Turiel, E. 1991. Children's reasoning about interpersonal and moral conflicts. *Child Development,* 62: 629–644.

Smith, G. A. 1992. *Education and the environment: Learning to live with limits.* Albany: State University of New York Press.

Snarey, J. 1985. The cross-cultural universality of social-moral development: A critical review of Kohlbergian research. *Psychological Bulletin,* 97: 202–232.

Sobel, D. 1993. *Children's special places.* Tucson: Zephyr Press.

Sobel, D. 1997. Mapmaking from the inside out: The cartography of childhood. *Orion Afield* 2(1): 14–19.

Solomon, R. P. 1992. *Black resistance in high school: Forging a separatist culture.* New York: State University of New York Press.

Soule, M. E. 1995. The social siege of nature. In M. E. Soule and G. Lease (Eds.), *Reinventing nature?: Responses to postmodern deconstruction* (pp. 137–170). Washington, DC: Island Press.

Spiegel, M. 1988. *The dreaded comparison: Human and animal slavery.* New York: Mirror Books.

Spiro, M. E. 1986. Cultural relativism and the future of anthropology. *Cultural Anthropology,* 1: 259–286.

Stack, C. 1996. *Call to home: African Americans reclaim the rural South.* New York: Basic.

Staub, E. 1971. A child in distress: the influence of nurturance and modeling on children's attempts to help. *Developmental Psychology,* 5: 124–132.

Stevenson, R. B. 1987. Schooling and environmental education: Contradictions in purpose and practice. In I. Robottom (Ed.), *Environmental education: Practice and possibility* (pp. 69–82). Victoria, Australia: Deakin University Press.

Stone, C. D. [1974] 1986. Should trees have standing?: Toward legal rights for natural objects. In D. VanDeVeer and C. Pierce (Eds.), *People, penguins, and plastic trees* (pp. 83–96). Reprint, Belmont, CA: Wadsworth.

Strong, D. 1995. *Crazy mountains: Learning from wilderness to weigh technology.* New York: State University of New York Press.

Tanner, T. 1980. Significant life experiences: A new research area in environmental education. *Journal of Environmental Education,* 11(4): 20–24.

Tarnopol, D., and Kahn, P. H., Jr. 1996. Divergent viewpoints on animal rights. Unpublished interview with D. Ringler.

Tavris, C. 1993. Beware the incest-survivor machine. *New York Times Book Review,* January 3, pp. 1, 16–17.

Taylor, P. W. 1986. *Respect for nature: A theory of environmental ethics.* Princeton, NJ: Princeton University Press.

Taylor, D. E. 1989. Blacks and the environment: Toward an explanation of the concern and action gap between blacks and whites. *Environment and Behavior,* 21: 175–205.

Thomashow, M. 1995. *Ecological Identity.* Cambridge, MA: MIT Press.

Thompson, B. 1996. AERA editorial policies regarding statistical significance testing: Three suggested reforms. *Educational Researcher,* 25(2): 26–30.

Thompson, J. 1995. Aesthetics and the value of nature. *Environmental Ethics,* 17: 291–305.

Thorkildsen, T. A. 1989. Pluralism in children's reasoning about social justice. *Child Development,* 60: 965–972.

Tisak, M. S. 1986. Children's conceptions of parental authority. *Child Development,* 57: 166–176.

Tisak, M. S. 1993. Preschool children's judgments of moral and personal events involving physical harm and property damage. *Merrill-Palmer Quarterly,* 39: 375–390.

Tisak, M. S. 1995. Domains of social reasoning and beyond. In R. Vasta (Ed.), *Annals of Child Development,* Vol. 11 (pp. 95–130). London: Jessica Kingsley.

Tisak, M. S., and Ford, M. E. 1986. Children's conceptions of interpersonal events. *Merrill-Palmer Quarterly,* 32: 291–306.

Tisak, M. S., Tisak, J., and Rogers, M. J. 1994. Adolescents' reasoning about authority and friendship relations in the context of drug usage. *Journal of Adolescence,* 17: 265–282.

Tisak, M. S., and Turiel, E. 1988. Variation in seriousness of transgressions and children's moral and conventional concepts. *Developmental Psychology,* 24: 352–357.

Trivers, R. L. 1971. The evolution of reciprocal altruism. *Quarterly Review of Biology,* 46: 35–57.

Turiel, E. 1983. *The development of social knowledge.* Cambridge: Cambridge University Press.

Turiel, E. 1989. Multifaceted social reasoning and educating for character, culture, and development. In L. Nucci (Ed.), *Moral development and character education: A dialogue* (pp. 161–182). Berkeley: McCutchan.

Turiel, E. 1998. Moral development. In N. Eisenberg (Ed.), *Social, Emotional, and Personality Development* (pp. 863–932). Vol. 3 of W. Damon (Ed.) *Handbook of child psychology.* 5th. ed. New York: Wiley.

Turiel, E. (in preparation). *The relations between children's social judgments and action.* University of California, Berkeley.

Turiel, E., Hildebrandt, C., and Wainryb, C. 1991. Judging social issues: Difficulties, inconsistencies, and consistencies. *Monographs of the Society for Research in Child Development,* 56(2, Serial No. 224).

Turiel, E., Killen, M., and Helwig, C. C. 1987. Morality: Its structure, functions and vagaries. In J. Kagan and S. Lamb (Eds.), *The emergence of morality in young children* (pp. 155–244). Chicago: University of Chicago Press.

Turiel, E., and Smetana, J. 1984. Social knowledge and action: The coordination of domains. In W. M. Kurtines and J. L Gewirtz (Eds.), *Morality, moral behavior, and moral development: Basic issues in theory and research* (pp. 261–282). New York: Wiley.

Turiel, E., and Wainryb, C. 1994. Social reasoning and the varieties of social experiences in cultural contexts. In H. W. Reese (Ed.), *Advances in child development and behavior,* Vol. 25 (pp. 289–326). Orlando: Academic Press.

Turner, J. 1996. *The abstract wild.* Tucson: University of Arizona Press.

Ulrich, R. S. 1984. View through a window may influence recovery from surgery. *Science,* 224: 420–421.

Ulrich, R. S. 1993. Biophilia, biophobia, and natural landscapes. In S. R. Kellert and E. O. Wilson (Eds.), *The biophilia hypothesis* (pp. 73–137). Washington, DC: Island Press.

Ulrich, R. S., and Lunden, O. 1990. Effects of nature and abstract pictures on patients recovering from open heart surgery. Paper presented at the International Congress of Behavioral Medicine, Uppsala, Sweden. June.

Ulrich, R. S., Simons, R. F., Losito, B. D., Fiorito, E., Miles, M. A., and Zelson, M. 1991. Stress recovery during exposure to natural and urban environments. *Journal of Environmental Psychology,* 11: 201–230.

Urmson, J. O. 1958. Saints and heroes. In A. I. Melden (Ed.), *Essays in moral philosophy* (pp. 198–216). Seattle: University of Washington Press.

van der Post, L. [1958] 1986. *The lost world of the Kalahari.* Reprint, New York: Harcourt Brace Jovanovich.

Vygotsky, L. S. [1934] 1962. *Thought and language.* Reprint, Cambridge, MA: MIT Press.

Vygotsky, L. S. 1978. *Mind in society.* Cambridge, MA: Harvard University Press.

Wainryb, C. 1991. Understanding differences in moral judgments: The role of informational assumptions. *Child Development,* 62: 840–851.

Wainryb, C. 1993. The application of moral judgments to other cultures: Relativism and universality. *Child Development,* 64: 924–933.

Wainryb, C. 1995. Reasoning about social conflicts in different cultures: Druze and Jewish children in Israel. *Child Development,* 66: 390–401.

Wainryb, C. 1997. The mismeasure of diversity: Reflections on the study of cross-cultural differences. In H. Saltzstein (Ed.), *Culture as a context for moral development: New perspectives on the particular and the universal* (pp. 51–65). New Directions for Child Development (W. Damon, Series Editor) San Francisco: Jossey-Bass.

Walberg, H. J. and Wynne, E. A. 1989. Character education: Toward a preliminary consensus. In L. Nucci (Ed.), *Moral development and character education: A dialogue* (pp. 37–50). Berkeley: McCutchan.

Walker, L. J. 1984. Sex differences in the development of moral reasoning: A critical review. *Child Development,* 55: 677–691.

Walker, L. J. 1986. Sex differences in the development of moral reasoning: A rejoinder to Baumrind. *Child Development,* 57: 522–526.

Walker, L. J. 1989. A longitudinal study of moral reasoning. *Child Development,* 60: 157–166.

Walker, L. J. 1991. Sex differences in moral reasoning. In W. M. Kurtines and J. L. Gewirtz (Eds.), *Handbook of moral behavior and development,* Vol. 2. (pp. 333–364). Hillsdale, NJ: Lawrence Erlbaum.

Watson, J. B. [1924] 1970. *Behaviorism.* New York: Norton.

Weiler, K., and Mitchell, C. (Eds.). 1992. *What schools can do.* Albany: State University of New York Press.

Wenz, P. S. 1995. Just garbage. In L. Westra and P. S. Wenz (Eds.), *Faces of environmental racism* (pp. 57–71). Lanham, MD: Rowman and Littlefield.

Westra, L. 1995. The faces of environmental racism: Titusville, Alabama, and BFI. In L. Westra and P. S. Wenz (Eds.), *Faces of environmental racism* (pp. 113–133). Lanham, MD: Rowman and Littlefield.

Wigginton, E. 1986. *Sometimes a shining moment: The Foxfire experience.* Garden City, NY: Anchor Books.

Williams, B. 1985. *Ethics and the limits of philosophy.* Cambridge, MA: Harvard University Press.

Wilson, E. O. 1975. *Sociobiology: The new synthesis.* Cambridge, MA: Harvard University Press.

Wilson, E. O. 1978. *On human nature*. Cambridge, MA: Harvard University Press.

Wilson, E. O. 1984. *Biophilia*. Cambridge, MA: Harvard University Press.

Wilson, E. O. 1992. *The diversity of life*. Cambridge, MA: Harvard University Press.

Wilson, E. O. 1993. Biophilia and the conservation ethic. In S. R. Kellert and E. O. Wilson (Eds.), *The biophilia hypothesis* (pp. 31–41). Washington, DC: Island Press.

Wilson, E. O. 1998. *Consilience: The unity of knowledge*. New York: Knopf.

Winters, W. G. 1993. *African American mothers and urban schools: The power of participation*. New York: Lexington Books.

Wohlwill, J. F. (1968). Amount of stimulus exploration and preference as differential functions of stimulus complexity. *Perception and Psychophysics*, 4: 307–312.

Wynne, E. A. 1979. The declining character of American youth. *American Educator: The Professional Journal of the American Federation of Teachers*, 3: 29–32.

Wynne, E. A. 1986. The great tradition in education: Transmitting moral values. *Educational Leadership*, 43: 4–9.

Wynne, E. A. 1989. Obedience and morality. *Ethics in Education*, 8(4): 2–4.

Wynne, E. A., and Ryan, K. 1993. *Reclaiming our schools: A handbook on teaching character, academics, and discipline*. New York: Macmillan.

Wyschogrod, E. 1990. *Saints and postmodernism: Revisioning moral philosophy*. Chicago: University of Chicago Press.

Youniss, J. 1980. *Parents and peers in social development*. Chicago: University of Chicago Press.

Index